THE ARCO BOOK OF
ELECTRONICS

Published by Arco Publishing, Inc.
215 Park Avenue South, New York, NY 10003

LIBRARY OF CONGRESS CATALOGUING IN PUBLICATION DATA
Main entry under title:
The Arco book of electronics.
        Includes index.
        1. Electronics – popular works. 2. Microcomputers –
popular works. 3. Television – popular works.
4. High fidelity sound systems – popular works.
5. Robots – popular works. I. Sturridge, Helena.
TK7819.A73   1984   621.381   84-2868
ISBN 0-668-06154-5

First published in 1984 by Kingfisher Books Limited
Elsley Court, 20-22 Great Titchfield Street, London W1P 7AD
A Grisewood & Dempsey Company

Typeset by Southern Positives and Negatives (SPAN)
Printed in Italy by Vallardi Industrie Grafiche, Milan

THE ARCO BOOK OF
ELECTRONICS

Helena Sturridge
Helen Mintern
Donald Aldous
Peter Marsh

ARCO PUBLISHING, INC.
NEW YORK

CONTENTS

MICRO-COMPUTERS

As part of the electronics revolution, microcomputers are transforming our lives. This introduction to microcomputers shows how they work, and looks at the range of devices that can be used with them. It also explains the binary system and looks at silicon chips, software and programming.

How Microcomputers Work

Silicon Chips

Input and Output Units

Software and Programming

Introducing Microcomputers

There are millions upon millions of computers in the world. Some are huge, cost small fortunes, and can tackle complex tasks such as which way the wind will blow tomorrow. Others cost less than 100 dollars, yet are still able to challenge you to a game involving a battle of wits.

Are computers the monsters of science fiction, poised ready to take over the world, or are they the fools that send gas bills to people living in all-electric houses? Neither portrait is particularly true.

Obedient Servants

Computers, once you get to know them, do what they are told. Their particular talents lie in their ability to process information. This means that if you give them the facts or figures, and tell them the rules by which they must work, then they will get on with the job.

In computer jargon, you would say that if you provide them with the data (facts and figures) and a program (the set of rules they must follow) then they will get on with the processing. If you give computers numbers to add, they will add them and do it faster and more accurately than any human ever could. If you would rather they helped you plan the best route to Mars, then tell them how to do it and they will do it with a thoroughness that few humans could match. There is no need to know anything about how microcomputers work to be able to use one. But if you want to find out what happens to your instructions, and how the data is used, this book tells you.

Micros, Minis and Mainframes

Computers come in all sizes. The largest are known as mainframes, and others are known as minicomputers. The smallest are micro-computers. They all process information, but the bigger the computer is, so it can handle more complicated problems, and do so faster. The difference is only one of scale.

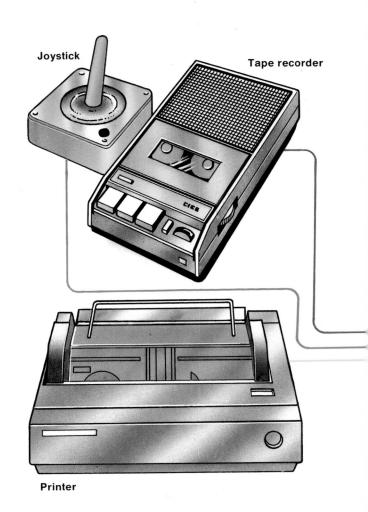

Joystick

Tape recorder

Printer

The Parts of a Microcomputer

When you see your first microcomputer there doesn't seem to be much to it – just a few boxes strung together. One looks like a typewriter keyboard and another looks like a television screen. There may be a cassette tape recorder plugged in at the back, or a disk drive. But don't be deceived by the simple appearance of a microcomputer. The outside gives no clues as to the machine's inner secrets.

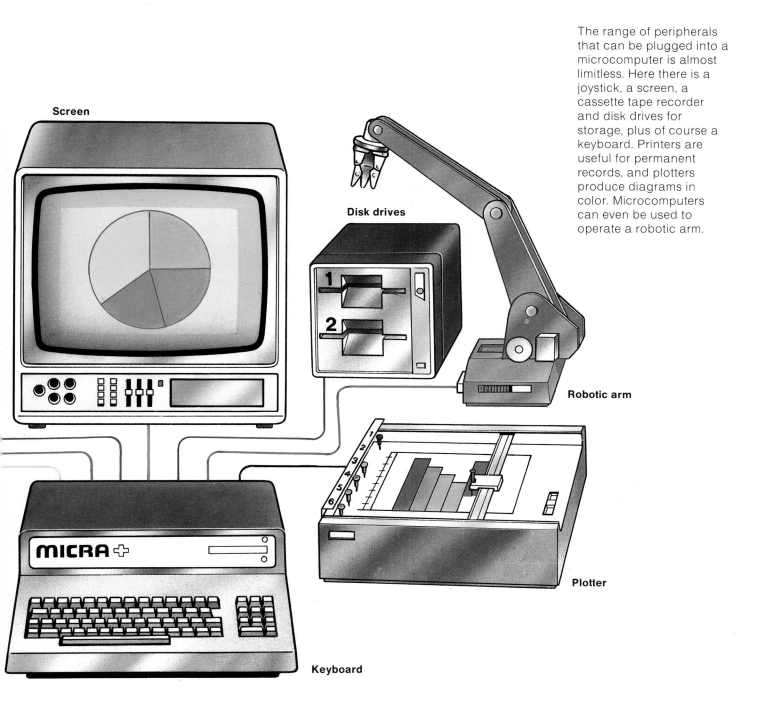

Screen

Disk drives

Robotic arm

Plotter

MICRA✛

Keyboard

The range of peripherals that can be plugged into a microcomputer is almost limitless. Here there is a joystick, a screen, a cassette tape recorder and disk drives for storage, plus of course a keyboard. Printers are useful for permanent records, and plotters produce diagrams in color. Microcomputers can even be used to operate a robotic arm.

A microcomputer's "brain", where all the information processing is done, is tucked away behind the keyboard. It is called the central processing unit, or CPU for short.

The keyboard and screen are the microcomputer's links with the outside world. Type your instructions in at the keyboard, and the answers will appear on the screen. These plug-in parts are the peripherals. The peripherals and the CPU are known as the hardware.

Software and Programs

To bring the hardware to life, it must be fed with precise instructions. Each group of instructions for a particular task is called a program. All the programs together make up what is known as the microcomputer's software.

It is the software that turns the basic machine into one specifically set up to play a game, or do a calculation. Each task must be tackled by a program.

Binary Codes and Logic

To get to know how a microcomputer works, it is essential to understand first how it handles information. The key lies in electricity. This is used as the basic building blocks from which a code is created. The code can be used to mean numbers, letters or just about anything. Morse Code uses lights flashing in a sequence of dots and dashes to mean, for instance, 'SOS'. Microcomputers use electricity that is either on or off. Particular sequences represent words or numbers.

On or Off: A Binary Code

In a way, the microcomputer's hardware has been designed to recognize and deal with two things – electricity that is ON or OFF. It has only these two options to distinguish between. They are the basis of its two-state, or binary, code.

In real life it is difficult to be sure that there is absolutely no electricity, or voltage, so most machines make do with the difference between higher and lower voltages. A higher voltage is read as on, a lower one as off. This is often written down as 0 and 1, where 0 means "off" and 1 means "on". Inside the microcomputer, everything is translated into a code made out of these binary 0s and 1s.

Bits and Bytes

Each individual 1 or 0, on or off, is called a bit. There isn't room for a code in a single bit. So, like Morse, the bits are grouped together. A group of eight bits is known as a byte (a byte is usually enough for a single letter of the alphabet). But there are no set rules, and different microcomputers prefer different-sized bundles of 0s and 1s. These bundles are called "words". Most microcomputers stick to byte-sized words.

Arithmetic

Once everything has been translated into the machine's binary-code language, it is still necessary to be able to work on it. Binary numbers can be added or subtracted, just as decimal numbers are. It may look a little odd, but the answers are the same. For example, $10 + 11 = 101$ in binary.

All you need to remember about binary code is that there are only two symbols, or digits: 0 and 1. With the decimal system, you work with ten digits from 0–9. In a binary system, every time you get past 1 you have to carry one into the next column, just as with decimals when you get past 9 and have to move into the next column.

◄ Microcomputers handle everything (numbers, letters, instructions) through electrical voltages. Imagine them as light-bulbs that are either on for a 1, or off for a 0. A row of light-bulbs would make a code according to which ones were on or off.

Binary and Decimal

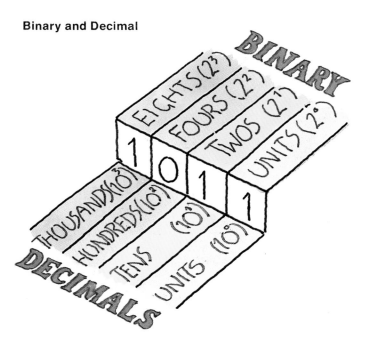

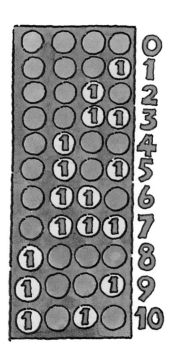

◀ Decimal numbers use ten digits from 0–9. Each column is equal to ten times the one before. So there are units, tens, hundreds and thousands. With binary, it goes up in twos since there are only two digits. So the columns mean units (2^0), twos (2^1), fours (2^2), eights (2^3) and so on. Numbers that look the same in both binary and decimal can mean very different things.

▶ Translation from binary to decimal is not difficult once you get the knack. Here is the translation for 0–10.

The Logical Approach: Moving House

You will have to go back if the previous owner has NOT moved out yet.

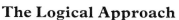

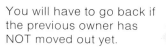

If the van is too tall OR too wide, you can go no farther.

If you have a van AND a driver, you can go.

The Logical Approach

But microcomputers don't stop with numbers – they also take decisions. They do this by following pre-set rules of logic called Boolean Algebra. These break decisions down into a series of yes/no questions. If certain conditions are met you get a YES, or TRUE. If not, you get a NO, or FALSE. This is another kind of binary code with only two options to choose between.

There are three main types of decision, known as AND, OR and NOT. Imagine you are trying to move house. To start, you need a van AND a driver to take the furniture. If you can say "yes" to both, off you go. On the way you must pass a low, narrow bridge. If your van is too high OR too wide, you will get stuck. Even when you get to the new house you will have to go back if the previous owner has NOT moved out yet.

The Working Machine

The presence and absence of electrical voltages, and the binary code, explain how a microcomputer holds on to facts and figures. But how does the machine deal with them? The working machine has three separate parts linked together. They are not always in different boxes, but they perform separate types of job.

Every microcomputer needs input and output devices to link the machine with the outside world. These might be a keyboard and a screen for example. Memories and storage devices are also essential to hold all the information that will be handled. The input, output and memory units copy, pass on and store the binary bits they receive.

It is only in the CPU, the central processing unit, that any actual work is done on the information. The CPU does the processing, adding or subtracting that is needed.

The diagram on the right explains how a CPU might step into action to add two numbers together. The machine must already have been loaded with a program to tell it how to add the numbers together, and you must feed in the two numbers at the keyboard.

Inside the CPU

The controlling force behind the CPU is the "clock" that sends out a regular signal millions of times every second. Against this the machine times and orders its every step.

The place to start is the program counter, one of many little memory spaces in the CPU that hold particular bits of information ready. The program counter always holds the "address" of the piece of information that will be needed next. In this example, it sends off to the main memory to collect the instruction. Its findings there are passed to the instruction register, and from there to the instruction decoder.

The instruction decoder prepares the circuits in the arithmetic and logic unit (ALU for short) for the addition. The ALU is where the actual processing is done.

The program counter, meanwhile, is ready with the "address" of the next item of information. It is the first number you have typed in at the keyboard. This is collected and put into the data register (since it is data rather than an instruction this time). The same process is repeated for the second number.

Buses

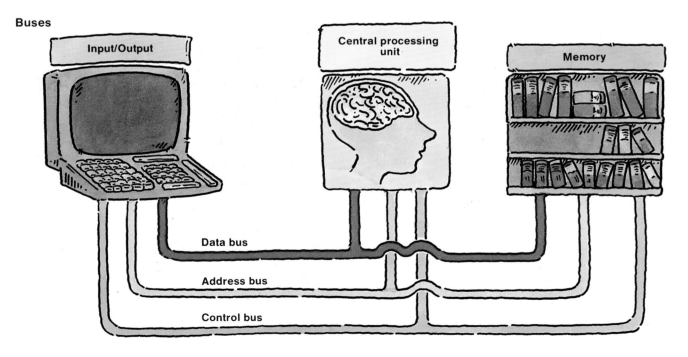

Input/Output

Central processing unit

Memory

Data bus

Address bus

Control bus

'Clock'

Program counter

Instruction decoder

Address register

Arithmetic and logic unit (ALU)

Accumulator

Instruction register

Data register

Keyboard

Memory

Screen

The Arithmetic and Logic Unit

So far, numbers have only been passed around and copied, never changed. Only in the ALU can they be processed. Once that is done (in this case they are added together), the results are passed to the accumulator. This is a temporary memory space, used while the machine waits to be told what to do with the answer (display it on the screen, for example).

Each stage of this calculation is handled in fractions of a second. The microcomputer follows the jobs it must do step by step. Each little binary bundle of information must be kept in order and tucked in the right corner of the memory ready to be found when the machine needs it.

◄ All microcomputers link their three parts with highways called buses, down which information can be passed.

▶ Nothing is forgotten. The memory stores every 0 or 1 in a system like marked pigeon-holes, each with its own address.

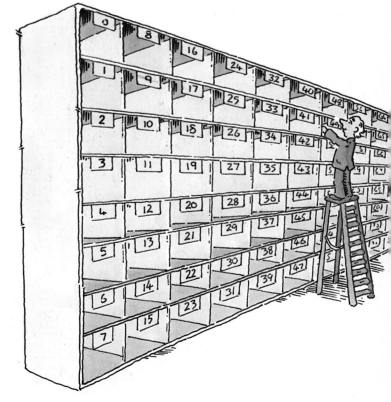

The Silicon Chip

You can't have much to do with a microcomputer without coming across silicon chips. These are the electronic circuits that store and process the information passing through the microcomputer in binary code. Chips come in various forms designed to do different things. Some store information and others process it.

Processing chips are needed in the CPU, where they apply the rules of Boolean algebra and arithmetic to working electrical circuits. This is done by building electrical switches that form what are called logic gates. The gates pass on, or change, the electrical 0s and 1s that go through them according to the required rules.

AND Gates

If, for example, the question is whether or not you can set off for your new house, then the conditions needed include the van and driver. With all AND gates, there must be room for two signals going in.

The first takes the answer to the question "Have you a van?" For a "yes", a 1 signal goes through. For a "no", a 0 signal goes through.

The second signal answers the question "Have you got a driver?" Again, there is a 1 for "yes" and a 0 for "no". The gate is set up to make sure that if two 1s go in, then a 1 will come out. This means that you can go off to the new house. If it were a 1 and a 0 (a "yes" and a "no") going in, then a 0 would come out. Without the driver (or the van), you couldn't leave.

OR and NOT Gates

OR gates work slightly differently. They will always produce a 1 ("yes") coming out if *either* of the two input channels has a "yes" going in. The NOT circuits have only one channel going in, and they automatically reverse the signal. If a 1 ("yes") goes in, then a 0 ("no") comes out.

Half Adders and Adders

The diagrams on the opposite page show that these AND, OR and NOT gates are combined to produce arithmetic circuits. The same electronic switches are used, but they are arranged with different results in mind.

Semiconductors and Silicon

Chips are made of silicon. Silicon is a semiconductor, which means that it is normally a poor conductor of electricity. But, if small amounts of other materials (such as boron and arsenic) are mixed with the silicon, it can be transformed into a good conductor. This is known as "doping" the silicon.

Silicon chips are made up of a variety of electrical components: transistors, capacitors and resistors. These are built into the surface of the silicon by doping patches of it to create the right electrical effects. With these it is possible to control the way the current flows, so building the gates required by the machine.

One-way Current

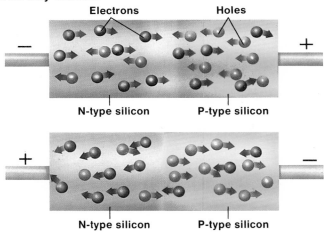

▲ The silicon in chips is doped to produce areas of positively charged (p-type) and negatively charged (n-type) silicon. P-type has some electrons missing, thus creating "holes", but n-type has spare electrons. In the top diagram, current flows. Electrons move across from the n-type toward the p-type silicon, and holes move from the p-type to the n-type silicon. In the bottom diagram, the current hardly moves at all because the holes and electrons are held back in their own areas. This means that current can flow in one direction only.

Half Adder

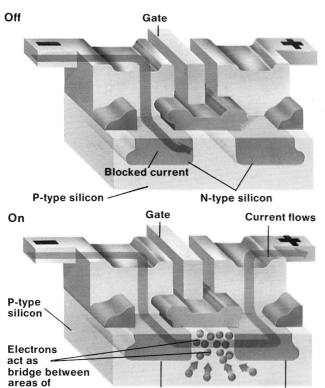

Full Adder

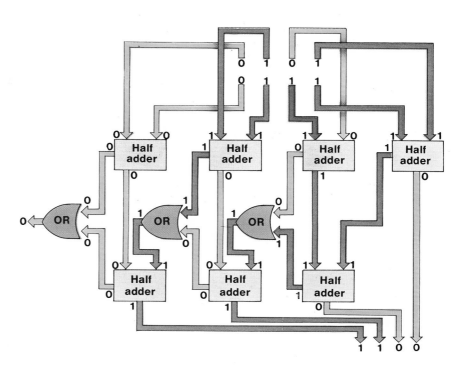

▲ The half adder will take two binary digits (in this case 1+1) and add them together, coping with any carrying that has to be done. To get the answer (10), the signal is passed through a series of gates.

▲ In this diagram of a full adder, two four-bit words (0101 and 0111) are added together. In this case, several half-adders are combined to produce the right results.

Off

Gate

Blocked current

P-type silicon

N-type silicon

On

Gate

Current flows

P-type silicon

Electrons act as bridge between areas of n-type silicon

N-type silicon

N-type silicon

◄ Silicon chips can control the flow of current by acting like switches, or transistors. In the top diagram the switch is 'off' and the current is blocked. At the bottom, the switch is 'on' and current flows.

▲ This is a small micro-computer on a single chip. It has thousands of circuits. Some act as memory spaces. Others are concerned with pro-cessing the information.

Making Silicon Chips

The first step in making a chip is to extract a tiny, pure silicon crystal. This is used to grow a huge one, by heating it in a furnace at very high temperatures. The rough-edged crystal is then polished into a smooth cylinder-shaped block.

The silicon block is sliced into wafers, each about four inches in diameter and less than a tenth of an inch thick. More than 250 separate chips are mapped out across the surface of each wafer.

Masking and Doping

Circuits are built up into the surface of the silicon by doping the pure crystal with impurities in the right places. This is done by using a series of photographic masks. On each of these is mapped out the exact pattern of the chips' circuits.

The masks are used with a photoresist. This is a plastic substance that hardens only under ultraviolet light. The wafer is coated with silicon dioxide, then soft photoresist, and the mask is laid on top. Ultraviolet light is shone on to the wafer. The parts exposed to the light through the mask are fixed in the photoresist as it hardens.

Under the mask, where there is no light, the photoresist stays soft. This is removed, leaving a pattern of hardened photoresist identical to the mask. The wafer is then bathed in acid, which eats away the exposed areas of silicon dioxide. The photoresist is removed, leaving the wafer marked out with the places where the first layer of doping must go. By using different masks, layer upon layer is built up on the wafer.

Testing

Once the layers have been laid down on the surface, the chips have to be tested. If one mask is slightly out of place, the layers won't connect properly.

The wafer is then chopped up into the

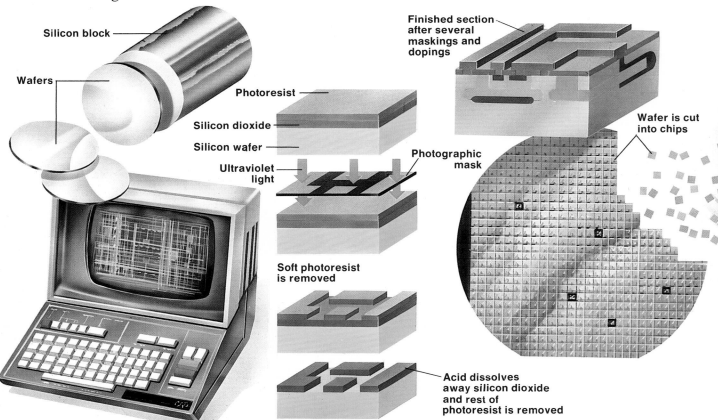

Silicon block

Wafers

Photoresist

Silicon dioxide

Silicon wafer

Ultraviolet light

Soft photoresist is removed

Acid dissolves away silicon dioxide and rest of photoresist is removed

Finished section after several maskings and dopings

Photographic mask

Wafer is cut into chips

individual chips. These are inspected with microscopes. Finally, the tiny chips of silicon are wired up with thin gold thread and embedded in plastic cases.

The Variety of Chips

There are hundreds of different kinds of chips. Complete microprocessors are probably the most complicated chips. They are designed to do the work of the CPU and so have logic gates, registers and memory spaces on their surfaces. Random-Access Memories, or RAMs, are designed to act as a microcomputer's main memory.

ROMs, or Read-Only Memories, are another variety. They are arranged to record and hold on to a particular pattern of 0s and 1s. Once programmed, a ROM will never forget the information it has been given. ROMs are useful for the basic programs in a machine that get used over and over again. Although most microcomputers are made from several chips linked together, it is possible to build a whole microcomputer (with CPU, RAM and ROM memories) on a single chip.

▲ The sausage-shaped block of silicon is sliced into wafers holding 250 or more chips on each. The chips are tested when they are finished.

▼ Everyone in a silicon chip factory must be dressed as if in a hospital to keep things clean. One speck of dust could ruin a chip.

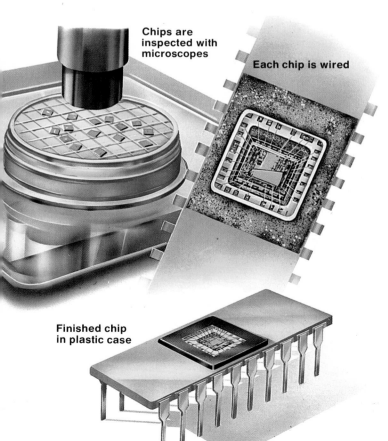

Chips are inspected with microscopes

Each chip is wired

Finished chip in plastic case

Storage Space

The main memory, made from a RAM chip, sits alongside the microcomputer's central processing unit. It holds all the information you might need to run a particular program, keeping the data and the program instructions ready for the central processor.

Main memories have disadvantages. If you switch off the electricity, they forget everything. Also, although they may have enough room for the program you are using, they soon get filled up. Most microcomputer users can turn to cassette tapes, floppy disks or hard disks as storage devices. These can hold all your programs and data, with or without a supply of electricity.

As you have already seen, computers work in binary code. This is translated by them into electronic signals which are either "on" for a 1 or "off" for a 0. The storage devices must therefore be able to register these sequences of 0s and 1s, while keeping them in the right order and position.

Each storage device does this through magnetism. The disks and tapes are made of magnetically-coated plastic. The microcomputer magnetizes tiny spots on their surfaces. If the spots are magnetized one way, they mean 0s. If they are magnetized the other way, they mean 1s. The spots remain unchanged until something is done to magnetize them in a different direction.

Keeping it on Tape

Tapes are the cheapest way of storing your programs and data. The magnetic spots are laid down in rows along the tape. They are most useful when the main memory is big enough to hold the programs in use, and you need somewhere to store information when the machine is switched off.

Tapes do have drawbacks if you want to use bigger programs or want to do more complicated tasks. They work serially: you have to start at the beginning of the tape and work

▼ Floppy disks come in various sizes and can hold as much as 360 kilobytes of information.

Floppy Disks

▼ The floppy disk spins inside the disk drive while the disk head hovers over the window ready to pick up the information.

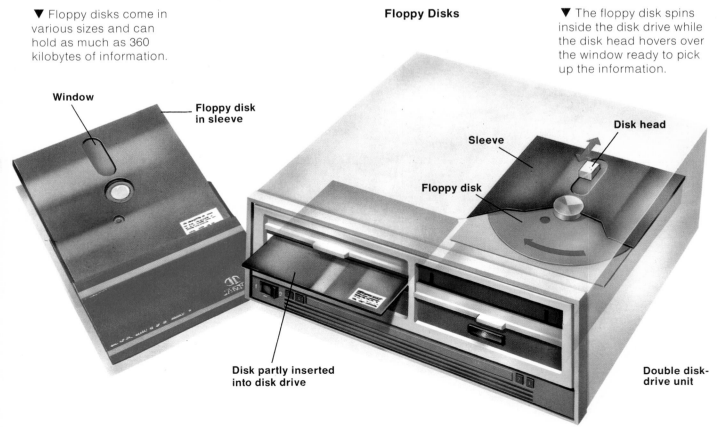

Window

Floppy disk in sleeve

Disk head

Sleeve

Floppy disk

Disk partly inserted into disk drive

Double disk-drive unit

your way through. This is fine for recording music, but not so useful with microcomputers if the bits of information you want are dotted about the tape.

The other problem with tapes is that cassette tape recorders sometimes don't have the accuracy that microcomputers need. A blurred binary 0 or 1 will confuse the microcomputer.

Floppy Disks and Hard Disks

Floppy disks are made of the same magnetically-coated plastic, and look like 45 rpm records permanently stuck in their sleeves. The surface of each disk is marked with circular tracks. The magnetic spots are laid down in these by the disk head, which hovers over the window in the sleeve until the right spot is underneath.

Fixed or hard disks, unlike floppy disks, are stacked up and sealed into boxes. They also have the nickname "Mini-Winnies" because they are smaller versions of the Winchester disk drives that were invented for big computers. The same type of tracks are laid down on the surface as in floppy disks, but hard disks are faster and can hold far more information than floppy disks.

Tape

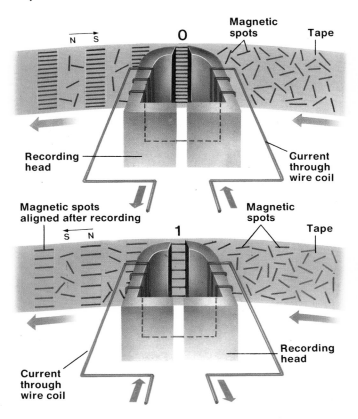

▲ Tapes store information as tiny magnetized spots. These are laid down on the surface of the tape by the cassette player's recording head. Current flows through the head in one direction to produce a 0 (top). It flows in the opposite direction to produce a 1 (bottom).

Hard Disks

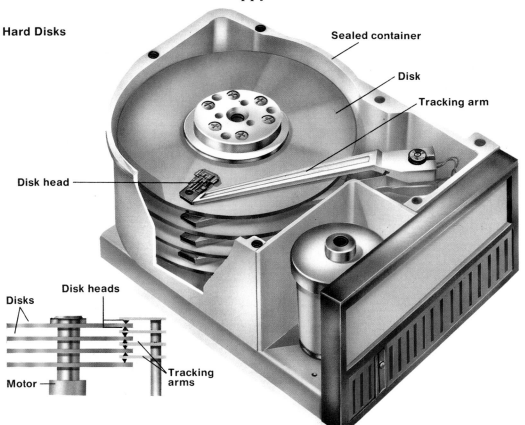

◄ Hard disks can hold up to 30 million bytes of information. The disks are sealed into their boxes and run faster than floppy disks. Separate heads hover over the surface of each disk.

Input Units

The important parts of the microcomputer, the central processing unit and the main memory, are tiny. They fit on a small printed circuit board. Most of the bulk of the microcomputer comes with the screen and the keyboard, the input and output units. Keyboards, for instance, have to be built to match the scale of human hands and fingers.

The Working Keyboard

The keyboard is usually laid out to look like a typewriter, so it is easy for people who are used to typewriters to know where the letters are. But microcomputer keyboards are more complex. The machine needs to know, for instance, when you are typing an ordinary word like "print" in a sentence about "this print is too small", and when you mean "Print" as an instruction.

So, the normal typewriter letters and numbers are supplemented by special keys to help you give instructions to the machine. There is no reason why keyboards shouldn't have a different button for everything you might want to do, but they would get very big and confusing, so instead they use combinations of keys. One combination, for example, will move the keyboard out of typing ordinary letters into a graphics mode. This produces dots instead of letters on a screen, just as when you press the shift key on a typewriter everything moves into capital letters.

Letters to Binary Bits

Whatever you do on the keyboard is understood by the microcomputer in binary bits. Beneath the buttons is a grid of criss-crossing wires. As each button is pressed down, a metal contact underneath touches a cross point on the grid. The microcomputer works out where the wires cross. It knows that at this point it must understand a particular code of bits. Ordinary letters, numbers and commands are translated straight away into the language, or code, preferred by the machine.

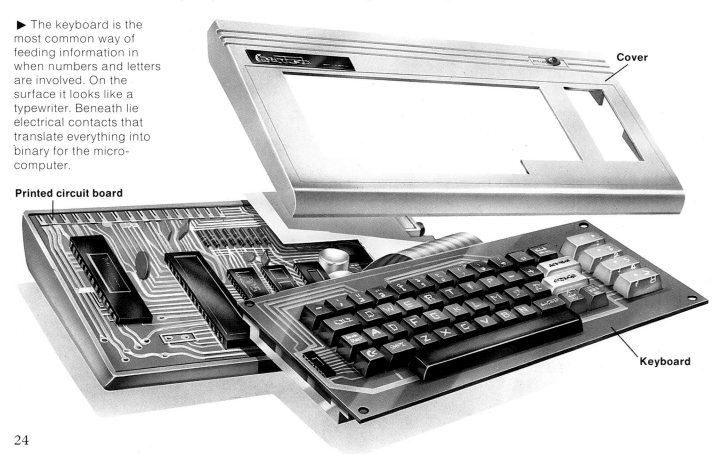

▶ The keyboard is the most common way of feeding information in when numbers and letters are involved. On the surface it looks like a typewriter. Beneath lie electrical contacts that translate everything into binary for the micro-computer.

Printed circuit board

Cover

Keyboard

Joystick

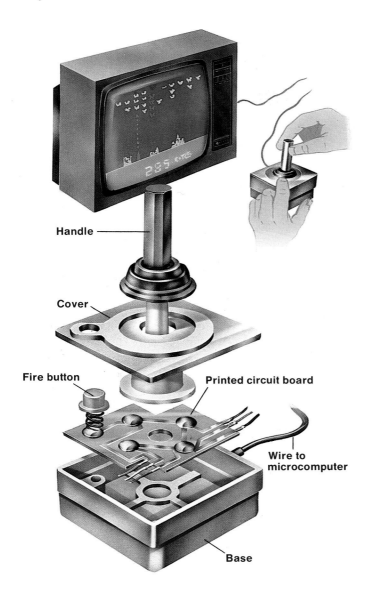

Handle

Cover

Fire button

Printed circuit board

Wire to microcomputer

Base

Lightpen

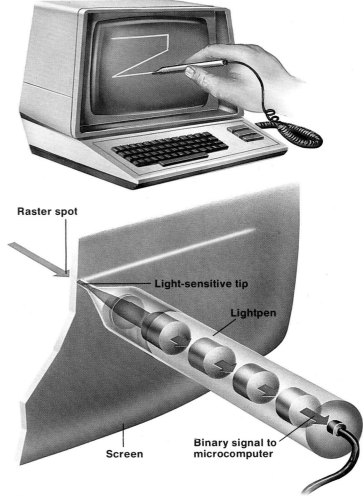

Raster spot

Light-sensitive tip

Lightpen

Screen

Binary signal to microcomputer

◀ Pushing the joystick handle forward brings two contacts together in the base. The microcomputer must be programmed to interpret this as a forward move.

▲ The tip of the lightpen recognizes the raster spot as it flashes past. At the same instant it sends a binary message to the microcomputer.

Joysticks and Lightpens

A keyboard is just one way of getting information into a microcomputer. When playing space invaders, it is much easier to guide your craft with something that reflects the movement on the screen. It is then that joysticks come in useful. As you move the arm around in its socket, the movements are translated into binary instructions that say up, left, down, as you shift about. When the time comes to fire, the button you press sets off the explosive pattern of dots.

The lightpen is another alternative to the keyboard. You can use it to answer questions, or to draw lines across the screen.

The lightpen works because a bright spot of light behind the screen alerts the pen's light-sensitive tip. The spot, known as the raster spot, moves in an orderly way across the back of the screen. Because of this, the microcomputer can work out where the spot was when the tip of the pen was alerted. Depending on how the microcomputer has been programmed, an action can then be set off.

More Input Units

It makes no difference to the microcomputer how it gets its information. But it is useful if it has an "input" port (a special socket) so that the input device can be plugged into the right part of the microcomputer's electronics.

Matching the Interface

It is also essential that the microcomputer and the input device use the same coding system. For example, it is no good if the device bleeps out a series of 1001110 to mean a full stop if the microcomputer uses a different combination of 0s and 1s to mean the same thing. The two machines must therefore have the right interface by understanding the same binary language.

If the microcomputer and input device do not understand the same language, however, it is often possible to use a translating device (programmed into a chip) which can be plugged into the machine.

Digitizing the Information

A microcomputer accepts all kinds of information. But it can't deal with any information unless you can help it to code it into 0s and 1s. Everything has to be "digitized", or turned into numbers. It is relatively easy to see how this is done with a keyboard, where the letters can just be coded straight away, but what about pictures?

To get a drawing into a microcomputer, it is necessary to have a digitizer. This looks like a pen and drawing pad linked to the microcomputer by a wire. But what it really is, underneath, is a grid. As with the keyboard, wires run up and down under the surface of the digitizer, crossing at specific places. As the pen crosses the grid, the microcomputer registers the individual spots where the grid wires cross.

In some cases it is the pressure of the pen that marks the spot to be registered. In others, you press a button on the end of the electronic pen to register the pen's position.

Using a Mouse with the Screen

With the device known as the "mouse", it is the electronics in the hand-held box that alert the grid beneath the digitizing board to the position that should be reflected on the screen. This gives you the ability to move a dot around the screen, as with a lightpen, but with the ease of a joystick.

▼ The 'mouse' has all the advantages of a lightpen but is as easy to use as a joystick. By pressing the buttons on the hand-operated box, you can move things around on the screen.

▼ Information on many packages now comes in the form of bar codes. A light beam measures the light reflected off the white bits between the stripes. The information is then fed to a microcomputer for processing.

The Mouse

Computer Chess

Communications

Telephone line

Acoustic couplers

Microcomputer's keyboard sends digital signal

Microcomputer's printer receives signal

◀ Many microcomputers can now recognize chess locations on a board. Given the right programming, they can beat you as well. The memory keeps a record of your moves.

▲ Microcomputers can be used to communicate via the telephone. Acoustic couplers turn digital signals into telephone signals at one end, and turn them back to digital signals at the other.

Bar code

Light-sensitive semiconductor

Signal to computer

Long-distance Information
Your keyboard can be many miles away from the central processing unit, provided that the digital signal can be bleeped to it without confusion. Microcomputers that are able to deal with communications use the telephone lines to do this.

The digital signal from the input device (for instance the keyboard) is sent down the telephone line and picked up at the other end by the microcomputer. Because telephones use a non-digital signal, an acoustic coupler, or a more sophisticated modem, is plugged between the machine and the line at both ends. The acoustic coupler translates digital signals into telephone-type signals at one end, and translates them back to digital signals at the other end.

Output Devices

Getting information into a microcomputer is all about turning the things that humans understand (such as letters and pictures) into digits that microcomputers can deal with. Getting it out again is just a matter of reversing the process. The most frequently used output devices are the screen and the printer.

Screens

Most screens used with microcomputers work along the same lines as television screens. In fact, many of the smaller home-microcomputers actually use a television screen. Others have specially-made screens which may give a slightly sharper image. These are often called visual display units, or VDUs. The electronics of a visual display unit are quite complicated, but it is worth remembering that it is still mainly a matter of handling 0s and 1s.

The inside of the screen is coated with dots of phosphor. These phosphors can be made to glow red, green or blue by firing electrons at them from behind. The electrons have to be guided to the right spot through the vertical and horizontal deflection systems. This ensures that the beam lights up the right phosphors to produce the correct image on the screen.

All the time, the screen deals with tiny points of light grouped together to form letters or numbers. It moves so fast that you never notice all the steps involved in putting together a screenful of information. To make it even faster, microcomputers often have a ROM (Read-Only Memory) programmed with the exact format of each letter, symbol and number.

The glow in the phosphors does not last long, so it is necessary to keep reminding the memory of what is on the screen. To do this, the memory keeps an exact reference of what is being shown on every patch of the screen. This is kept in 0s and 1s. When the screen image changes, so do the codes in the memory.

Dot Matrix Printer

Raster Scanning

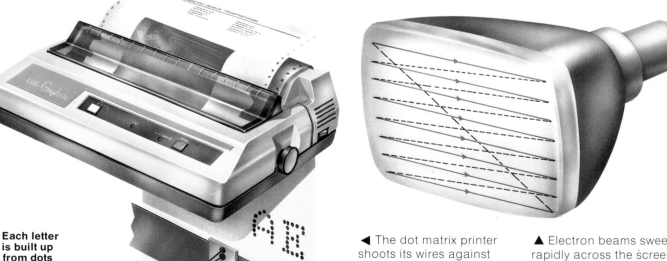

Each letter is built up from dots

Paper

Wires

Ribbon

Print head

◄ The dot matrix printer shoots its wires against the inky ribbon. It forms letters and numbers by building up a pattern of dots. It can work fast enough to print 120 characters a second. Pictures take a bit longer.

▲ Electron beams sweep rapidly across the screen in a series of lines to build up a picture. This is known as raster scanning. When the beams reach the bottom of the screen, they return diagonally to the top, ready for the next sweep.

How the Screen Works

Electron guns fire beams of electrons at the phosphor-coated screen. Each beam corresponds to one color: red green or blue. A deflection system, controlled by the micro-computer, guides the beams to the right points on the screen.

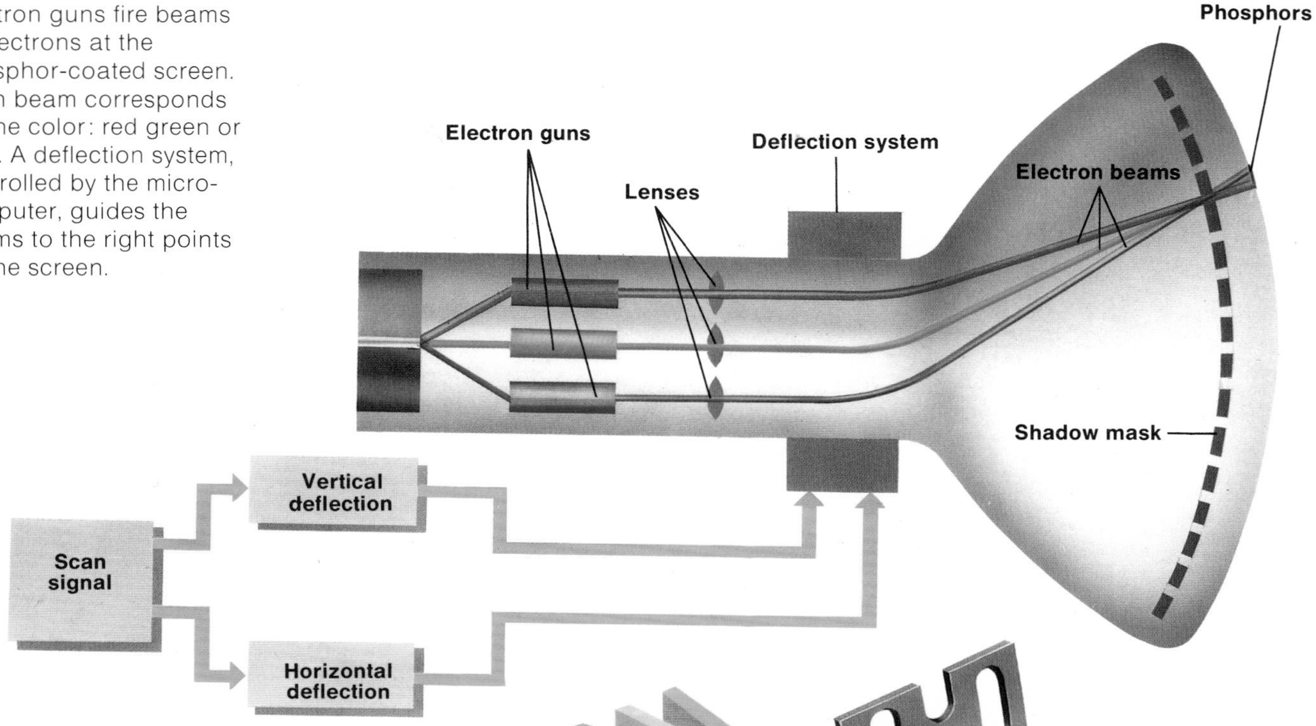

Electron guns

Lenses

Deflection system

Phosphors

Electron beams

Shadow mask

Scan signal

Vertical deflection

Horizontal deflection

▶ The shadow mask, which lies just behind the phosphor-coated glass, guides the beams to the phosphors corresponding to their color. The phosphors glow with color when they are struck by the beams.

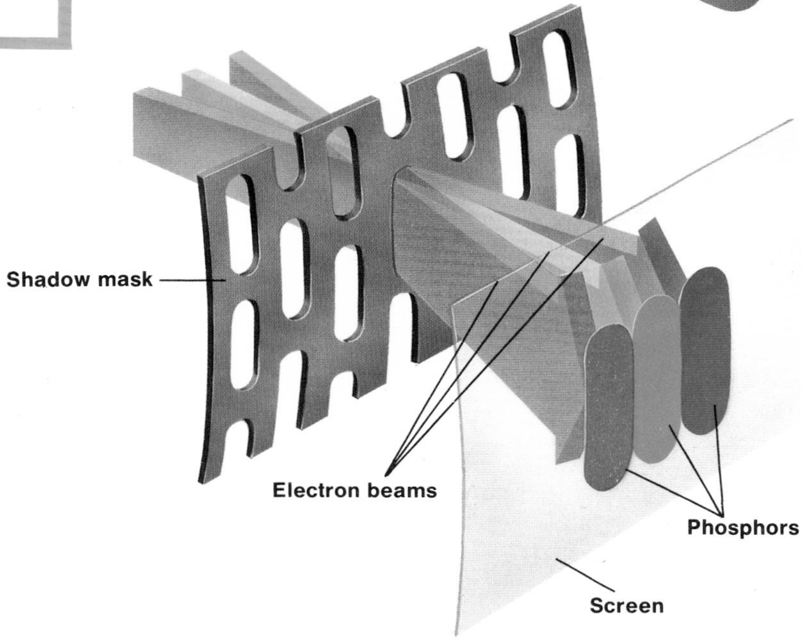

Shadow mask

Electron beams

Phosphors

Screen

Printers

Visual display units are convenient to work with, but they don't provide you with a permanent record as printers do. There are many different types of printers using different mechanisms. One of the most common used with microcomputers is the dot matrix printer.

Underneath the cover of a dot matrix printer is a series of wires lined up against a ribbon. The wires are held back from the ribbon by magnets. When the time comes to print, the wires are released. They smack into the ribbon, leaving an inky mark on the paper beneath. Several dots grouped together produce a letter or number.

This may sound slow and laborious, but the wires move at such a rate that they can produce 120 characters (letters or numbers) every second. A ROM chip usually holds the information needed to turn the computer's 0s and 1s into signals to the print head to make the right pattern of dots.

More Output Devices

A microcomputer can be linked to almost any output device, provided that they both use the same language. The interfaces must match, as with input devices.

Putting it Down on Paper

Dot matrix printers are only one kind of paper output. Thermal printers have wires that produce electric sparks to blacken the surface of heat-sensitive paper. This type of printer is quiet because it does not hit the paper hard. But the paper is more expensive because it has to be specially treated.

Daisy-wheel printers work on an entirely different principle. Beneath their covers they have a spinning wheel with each letter and symbol marked on the end of a spindle. When the right letter hovers over the page, a hammer comes down and slams the letter against the ribbon onto the paper. Daisy-wheel printers make sharper images on the page, but they cost more and print at a slower rate than dot matrix printers – around 60 characters per second. The bits (binary digits) that the microcomputer produces are turned into physical movements by the print head, which is often governed by a chip of its own.

Pictures on Paper

Printing words and numbers is a delicate operation, but printing pictures is more so. With letters and numbers, short cuts can be made because the letters are already known and familiar. With pictures, every dot on the page has to be separately accounted for, as each picture is different.

Plotters, for instance, will draw you a graph or map out the world by tapping pens up and down on a sheet of paper until every patch is

Speech Synthesizer: How it Works

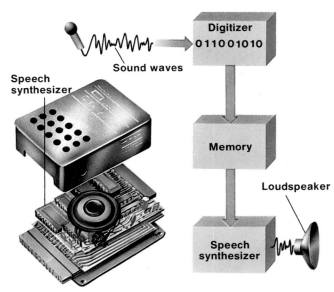

Daisy-wheel Printer

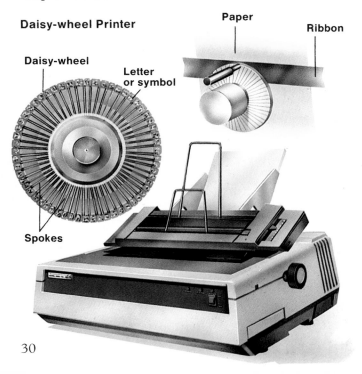

◀ The letters and symbols of the daisy-wheel lie at the end of each of the spokes. As the wheel spins, a hammer punches them against the ribbon at the right moment. If you want different styles of letters or unusual symbols, you can change the daisy-wheel.

▲ Sound waves are converted into digital form and stored in the micro-computer's memory. The speech synthesizer converts the digital signal back into sound waves.

▶ Plotters are useful when you want accurate drawings. Colored pens move over the surface of the paper according to the way the micro-computer has been programmed.

appropriately colored. The bar stretching across the paper moves horizontally, while the pen moves along the bar. Just as the digitizer turns every point on a line into a dot with specific co-ordinates that can be stored as binary digits, so the plotter turns those same digits back into the dots on a page.

Robotic Arms

It is only very recently that microcomputers have been able to use robotic arms. It takes a great deal of programming to instruct a hunk of metal how to move. Imagine all the tiny details involved in stretching out your arm to pick up something: the twisting and turning of your wrist and elbow, and the measuring of the distance. All this must be explained in detail to the microcomputer so that it can instruct a robotic arm to behave in the same way.

Speech Synthesizers

Microcomputers can also "talk" with the aid of speech synthesizers. The process is still dependent on 0s and 1s. When a voice is recorded, the component sounds of the words are broken down into a digital form and stored away in the computer's memory. When the time comes for the synthesizer to speak, the 0s and 1s are returned to a form that approximates to the speaking voice. Because people's voice patterns vary, this process requires detailed coding, so it takes up a lot of power and memory space.

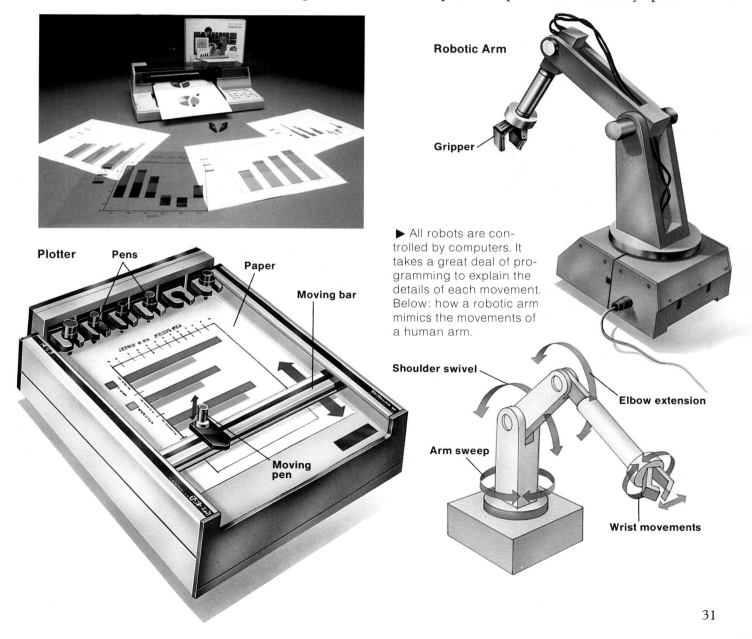

Robotic Arm

Gripper

Plotter   **Pens**   **Paper**

Moving bar

Moving pen

▶ All robots are controlled by computers. It takes a great deal of programming to explain the details of each movement. Below: how a robotic arm mimics the movements of a human arm.

Shoulder swivel

Elbow extension

Arm sweep

Wrist movements

Software

All the metal, plastic and silicon that make up a microcomputer's hardware would be useless without software. Hardware passes around the electronic code of 0s and 1s. It needs the software to arrange and order that code. It is the software that turns your microcomputer into a chess opponent rather than a history teacher, or a mathematician rather than an astrologer. With a full range of software, a single machine can do hundreds of different jobs.

The Guiding Rules

Software is the rules by which the machine is guided. Between them, the programs that make up the software do everything from telling the CPU how to locate the keyboard to displaying a game on the screen and beating you at it.

Software programs have to be very precise. They must take the microcomputer step by step through the job to be done in such detail that there is no room for confusion. It doesn't matter if the procedure is long-winded, providing the instructions are exact.

Systems Software

Software can be thought of as divided into two layers. The systems software represents the inner layer. This tells the microcomputer how to look after itself. It shows the machine everything, from how to use its memory to how to flash something on its screen. It guides the electronics through every step they must take. When you buy a microcomputer, the systems software comes with it. (It is also known as the operating system.) Without systems software, very little would happen when you switched on the machine.

Languages, Compilers and Interpreters

One of the most important items in the bundle of programs that make up the systems software is the compiler or interpreter. These are automatic translators that turn all the programs into the 0s and 1s that the machine must deal with. Human programmers find it difficult to think in 0s and 1s, so they use special computer languages.

▶ Applications software consists of all the programs you work or play with. These can range from computer games to educational and business programs. Each program is translated by the compiler or interpreter into the microcomputer's own code.

◀ There is a great variety of ready-made programs for microcomputers but it can be challenging to write your own.

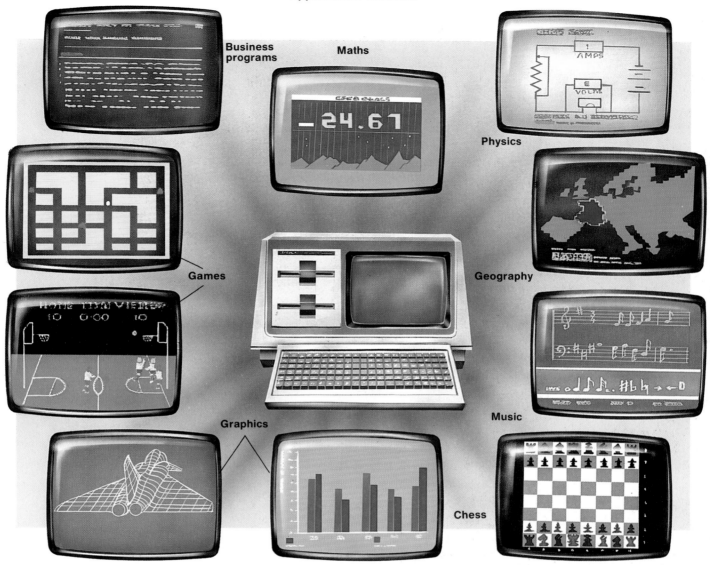

Business programs

Maths

Physics

Games

Geography

Music

Graphics

Chess

These languages are not as difficult to learn as French or German; they are like a shorthand for programmers. Well-known languages include Basic, Pascal, APL, Cobol and Fortran. Once a program has been written in one of these languages, it must then be translated by the compiler or interpreter into the code that the machine understands.

Compilers and interpreters work in slightly different ways. Compilers translate everything before they start. Interpreters do it line by line as they go through the program. They are a bit slower. But this difference is something that usually only matters to professional programmers.

Applications Programs

Applications programs represent the outer "layer" of software. These are the ones you buy, or write yourself, to play games or teach you French, for example. An applications program controls what you actually do with the microcomputer, while the systems software concentrates on how the machine behaves.

Writing Software

Software is difficult to write, but that is half the fun. Languages make it easier to write your own programs, but they don't take away the challenge. It takes good ideas that are clearly explained to produce a good program.

Programming

Programming requires thoroughness and great attention to detail. At times it can seem extremely pedantic as you work through the instructions, but when they come out right it all seems worthwhile. Few programs ever come out right first time. Mistakes in programs are known as bugs.

Drawing up a Flowchart

Imagine that you want a robot to produce breakfast for you every morning. First you must plan exactly what is involved in producing breakfast. This plan can then be drawn up into a flowchart (like the one on the right) which shows in diagrammatic form the steps your program will have to cover.

The first thing to do is get the robot to switch itself on at 7 am ready to go: START. Then it must lay the table, put the cornflakes in a bowl, add the milk and sugar and leave you to eat in peace. Meanwhile, it must get on with cooking the scrambled eggs, put them on a plate and serve the lot up. But first the robot must take away your empty cereal bowl. Finally it should go back to the beginning, ready to start again tomorrow morning.

Improving the Flowchart

This seems quite straightforward, but is it foolproof? In fact, you have scarcely started before you are up to your neck in cornflakes, drowning in milk and covered in sugar – you forgot to say how much you wanted. The robot has no idea what is a reasonable helping. So add a few measurements and try again.

This time you have hardly got the first spoonful of cornflakes to your mouth when the robot tries to whip the bowl away and dump the eggs and bacon on you. So you must make it stop and find out if you are ready. A diamond-shaped box signifies that a decision must be taken.

Things seem to be working quite smoothly now. But are you sure you are going to want this enormous breakfast every morning? Once it

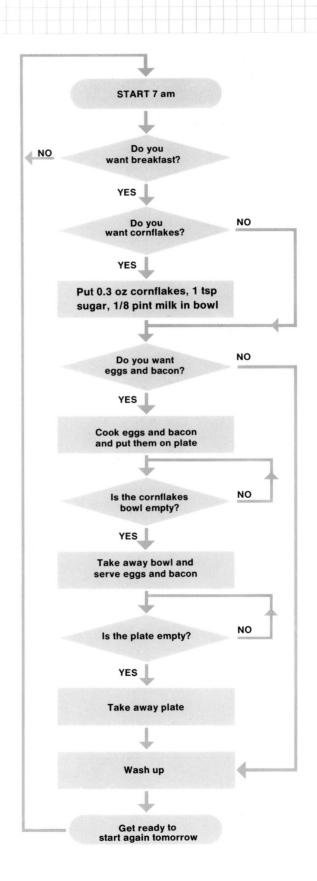

starts, the robot will go on for ever, and if you don't eat them the eggs and bacon will pile up day after day. So you had better leave yourself the option of getting out of breakfast if you want to.

This of course is just the start. You need to detail how the robot should lay the table, and what with. You also need to explain how to cook, and what to do when washing up. And, if you don't want to run out of cornflakes, you had better tell the robot to warn you when its supplies are low.

A Program to Try

This is a program for the Sinclair ZX81 which is the smallest and cheapest computer around. The program will draw stars or asterisks (✳) all over your screen. You can control the ✳ by pressing the 8 key to go right, the 7 and 6 key to go up and down, and the 5 key to go left.

As you will see, all program lines have to be numbered, so that the computer knows what order to follow them in. Although the numbers could be 1, 2, 3 and so on, it is usual to space them out in 10s so that if you forget to put in a line you can easily go back and squeeze it in. Microcomputers also like to have a stroke (/) through their zeros. This is so that there is no confusion with the letter O.

How the Program Works

Lines 10 to 30 just get the microcomputer started in the middle of the screen. X tells it how far across to go, Y how far up or down. Line 30 tells the machine what to put on the screen when it gets there.

Lines 40 to 70 turn the keys numbered 5, 6, 7 and 8 into control keys to move the ✳ around. If you press 5, you skip one space to the left. If you press 8, you move one to the right. By pressing 6, you move one space down, and by pressing 7 you move one space up.

The last line takes you back to line 30. Without line 80, the computer would stop when it gets to the bottom of the list. With "Goto", it knows to go back to line 30 and do whatever it is told. In this case, it puts stars all over your screen.

Program for the Sinclair ZX81

```
10   LET X = 15

20   LET Y = 10

30   PRINT AT Y, X; " ✳ "

40   IF INKEY$ = "5" THEN LET X = X − 1

50   IF INKEY$ = "6" THEN LET Y = Y + 1

60   IF INKEY$ = "7" THEN LET Y = Y − 1

70   IF INKEY$ = "8" THEN LET X = X + 1

80   GOTO 30
```

Working with Microcomputers

Microcomputers break the world down into simple elements. Every letter, number, color or sound must be reduced to the 0s and 1s of its binary code. Only then does careful ordering and speedy processing take the machine through the logical steps to find the correct result.

The Microcomputer's Skills

A microcomputer's talents can be put to use in five main areas, each of which calls on different aspects of its skills. *Data processing* comes in where information must be considered and decisions taken. At other times it is speedy *calculation* that is needed. *Text processing* is more a matter of storing and printing words than actually processing them, while *information retrieval* makes use of the microcomputer's reliable memory. But the machine can also be used to *control* other pieces of equipment by switching them on and off, and monitoring them automatically.

▲ Microcomputers can process data, handle words and carry out calculations. They are at their best when they are called on to use all their skills. In this way, your microcomputer can act as your personal tutor.

▼ Microcomputers can be used to teach you music. They can help you to understand rhythm and harmony, and it can be great fun to listen to your own musical compositions as you write them.

A Microcomputer in School

All these talents could be put to use in your school, provided your machine had the right programs to suit the jobs. Organizing the school timetable would be a good way of employing the machine's data-processing talents. Teachers, classrooms, books and children could be matched up in a fraction of the time it would take the headteacher to consider the possibilities.

The school newspaper could make use of text-processing. Storing articles, correcting spelling and printing copies would all be easy jobs for the machine.

The microcomputer's mathematical skills might come in useful for the school accounts. Costing out meals and paying the teachers' salaries would be done with concise accuracy.

Information-retrieval comes in handy in the library. The catalog of books, if stored on a microcomputer, would take up a lot less space than a card file, and could be much easier to search through.

Microcomputers are used for controlling equipment in dozens of ways. One school's machine "listens" out for the noise of pupils arriving in the morning and uses it to trigger off the switch that turns on the heating system. The quiet that reigns when everyone goes home makes the machine switch the heating off at night.

Teaching the Microcomputer to Teach

The great advantage of the microcomputer is that it can let you try things out. For example, a principle in physics can be demonstrated and then tried out on the machine. When things go wrong, the machine can take you back through the theory and explain it again. It is far easier to test your skills in bridge-building with a microcomputer than to wait until you have tried to build one that has collapsed.

But this is just a beginning. People come up with new ways of using microcomputers all the time. You don't have to be a university professor to write a good software program. More often than not, the best games are written by the people who want to play them.

▲ With their vastly expandable memories, all computers make good electronic libraries. Prestel is one of the best-known. Through your television screen you can gain access to pages of data.

▼ Learning languages can be as exciting as playing a space invaders game. Well-designed programs don't just drill you on irregular verbs, but invite you to pit your wits against the microcomputer.

TELEVISION
AND
VIDEO

Television and video have become an essential part of our lives – in entertainment, education and even security. This section shows how television programs are made, transmitted and received. It also looks at video recorders and video disks and gives hints on using video cameras.

How Television Works

Making Television Programs

Video Recorders and Video Disks

Making a Video

The Scope of TV and Video

When you are grown-up and your children are going to school, this book may not exist. In fact, schools as you know them may not exist either, and libraries with books may be museums. All this will happen because of television and video. Television was invented during the 1920s by John Logie Baird.

Studying by Television

Let's visit a home of the future, say in the year 2000, and see what everyone is doing. Alice is 12 years old. She is not wasting time watching television; she is at school. That's her teacher on the screen. She manages to see Alice once a week to check her written work, but not for long. By teaching on television she could have a thousand pupils in her class at once, but she doesn't have more than a hundred. Alice likes to "go to school" in the living room where there is a row of flat screens against the wall. She wears headphones to listen to her teacher.

Alice's brother, Peter, likes to work by himself in his bedroom with a smaller, personal screen. He is 20 years old and, although he lives in California, he's studying with the Massachusetts Institute of Technology. His microcomputer and screen are linked by telephone to the local library. They are sending Peter a new article written in Cambridge. They received it overnight from Massachusetts during the cheaper-rate computer time.

Work and Leisure

Father has worked at home for the last five years, ever since his supermarket became fully automated. As supply manager for the supermarket, he checks the stock on the shelves every Monday morning visually through the closed circuit television cameras. On his home screen he can also study the computer totals produced by the automatic checkout tills.

Mom is watching a live television program. Her favorite daytime program is the 24-hour European news station which the family receive through their satellite dish receiver on the roof.

Grandad Jones is the only member of the family who uses the video disk. At the moment he's looking at a dahlia catalog, and the video disk gives the best picture available on any system.

Grandmother Jones can hardly walk and spends her time watching the goings-on out in the street through the local closed circuit camera system. The council originally set up the system to help stop burglaries.

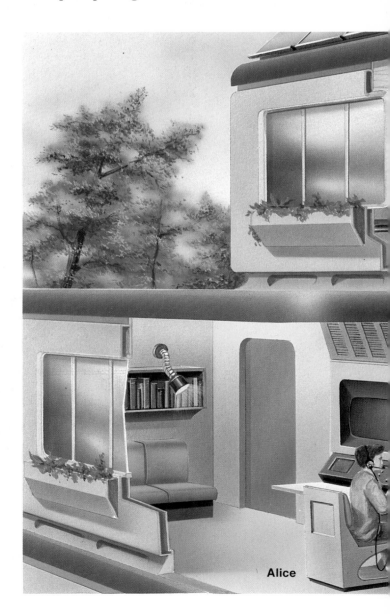

Alice

40

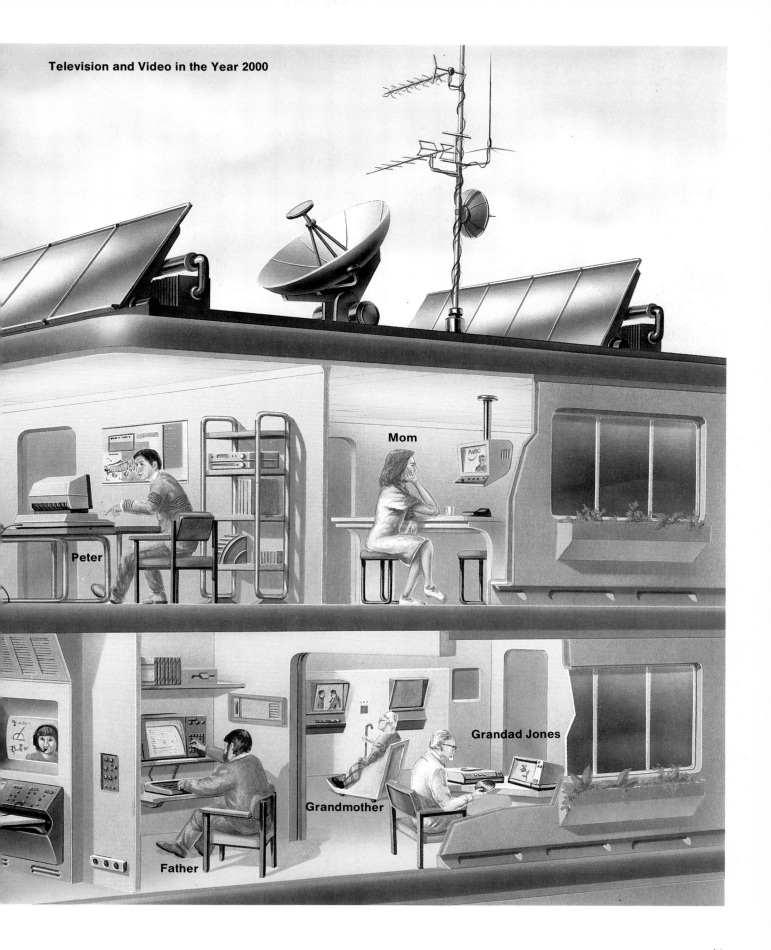

The TV Receiver

Whatever else you may have in your home in the year 2000, you can be sure there will be some kind of television set. The word *television* actually means "pictures from far away", but to understand how the set produces these pictures it might be better to use the word "dottyvision" instead.

How Television Works

Your television set has three main parts: the tuner, the loudspeaker and the cathode ray tube, as well as various electronic parts which deal with the signal for the cathode ray tube.

The tuner receives different signals from several television stations through the antenna, which picks the signals out of the air. The tuner separates the pictures from the sound. These have traveled together from the transmitter (see pages 44–45). The part of the tuner you can see is a set of knobs or buttons which enable you to tune in to each channel.

The loudspeaker receives the sound as waves of electronic signals. The signals activate a magnet in the loudspeaker. In turn, the magnet makes a cardboard cone vibrate many times a second, disturbing the air and producing the noises we hear.

Receiving Color Pictures

The picture signal is sent to the cathode ray tube. You will recognize the wide end of this as the television screen. The screen is coated inside with chemical phosphor dots. These are arranged in sets of three to produce the three different colors of television – red, blue and green. Behind the screen is a thin plate called a mask with vertical slots cut in it. Each slot is for one group of dots.

The narrow end of the cathode ray tube is hidden by the casing of the set. It contains three electron guns. The guns spray electrons on to the phosphor dots through the slots in the mask. The phosphor dots glow with color to form a pattern, or picture, when hit by the electrons.

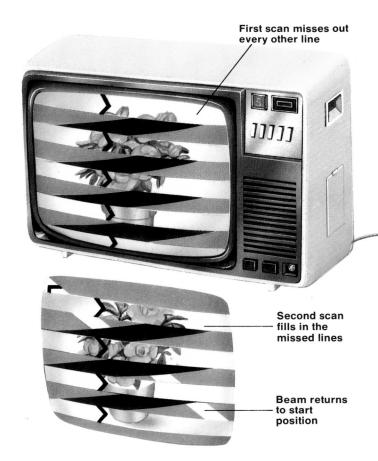

First scan misses out every other line

Second scan fills in the missed lines

Beam returns to start position

▲ The picture looks steady, but the electron beams scan the screen 50 times each second. The phosphor dots glow red, blue or green to produce the pattern of each picture. If you go very close to the screen you can see the dots.

▶ The main part of the TV set is the cathode ray tube. The big end is the screen you see. The narrow end contains three electron guns. These spray electrons through the shadow mask on to the phosphor dots. Some sets use only one gun.

Each dot can only receive electrons from one gun because of the way the slots are positioned. Some of the newest sets have only one gun to spray all three colors.

The spraying action is like a garden sprinkler, zig-zagging horizontally across the screen in lines. In Europe, television pictures have 625 lines and in the United States they have 525. The guns carry out two scans. The first misses out every other line. The second fills in the

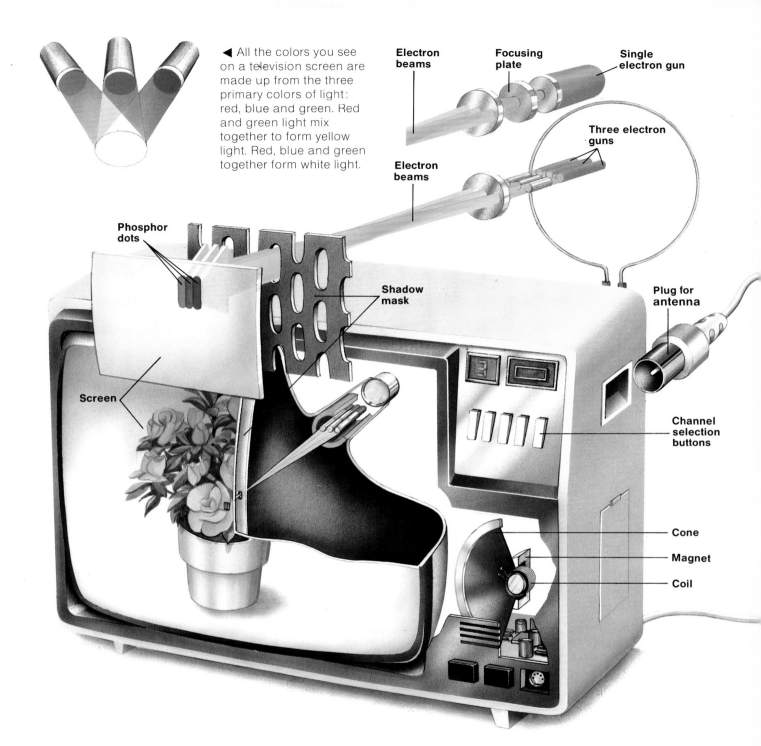

All the colors you see on a television screen are made up from the three primary colors of light: red, blue and green. Red and green light mix together to form yellow light. Red, blue and green together form white light.

Electron beams

Focusing plate

Single electron gun

Three electron guns

Electron beams

Plug for antenna

Phosphor dots

Shadow mask

Screen

Channel selection buttons

Cone

Magnet

Coil

missed lines. All this happens 50 times every second.

Black and White Pictures

The same phosphor dots and electron guns are used to produce black and white pictures. The three colors are mixed to make black, white and gray shades. It may be difficult to understand how this is possible because we are used to mixing paint colors, which behave differently from the colors in light.

If you want to experiment with mixing light, try this. Take three torches and put red film in front of one, blue film in front of another, and green film in front of the third. Now shine the red, green and blue lights on to a white screen. The area where all three colors overlap will be white.

Pictures from the Air

Like the simplest pocket camera, the television camera has an "eye", or lens. Light from objects enters the camera through the lens and is changed into electrical signals. These are combined with carrier waves and sent out into the air from transmitters. The television set at home picks up these signals and converts them into a picture on the screen.

How the Camera Works

Inside the camera are three tubes, and at the end of each tube is a target plate. These plates are coated with a light-sensitive material which alters when touched by light. All objects reflect light. Light coming from the subject enters through the lens of the camera. It touches the target plates and the light-sensitive material releases electrons which move down the tube.

An electron is part of an atom. When electrons start breaking away from the atoms on the target plate, they leave behind a pattern of positive electrical charges. These charges contain all the information needed to produce a picture on a television screen.

Once this pattern of information has been created, it needs to be carried out of the camera. This is done by using electron guns, which are situated at the other end of the three tubes. Each gun fires electrons at its target plate in order to read, or scan, the pattern and turn it into electrical signals. The burst of electrons scans the target plate in a similar way to the electron guns inside the television set, and is also very fast. The target is scanned 50 times every second.

Color Television

Before the arrival of color television, the camera had only one tube. Now that colors are being transmitted, light coming through the camera lens has to be split up into its three primary colors: red, blue and green.

Each of the three camera tubes handles one color. Light that looks white to us is actually

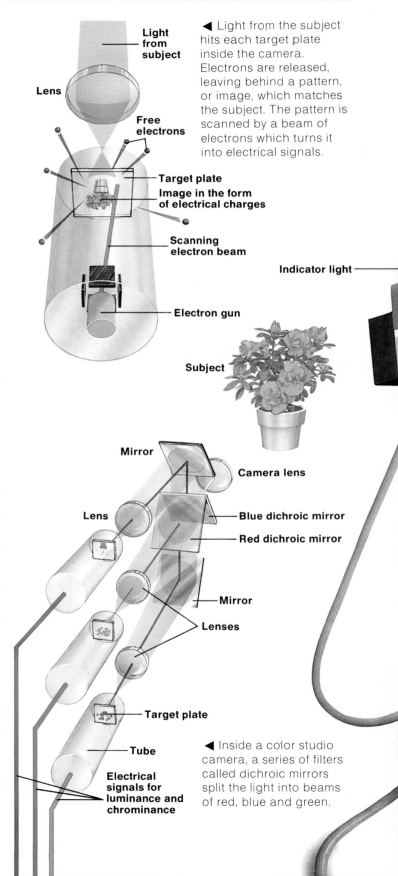

Light from subject

Lens

Free electrons

Target plate

Image in the form of electrical charges

Scanning electron beam

Electron gun

◀ Light from the subject hits each target plate inside the camera. Electrons are released, leaving behind a pattern, or image, which matches the subject. The pattern is scanned by a beam of electrons which turns it into electrical signals.

Indicator light

Subject

Mirror

Camera lens

Lens

Blue dichroic mirror

Red dichroic mirror

Mirror

Lenses

Target plate

Tube

Electrical signals for luminance and chrominance

◀ Inside a color studio camera, a series of filters called dichroic mirrors split the light into beams of red, blue and green.

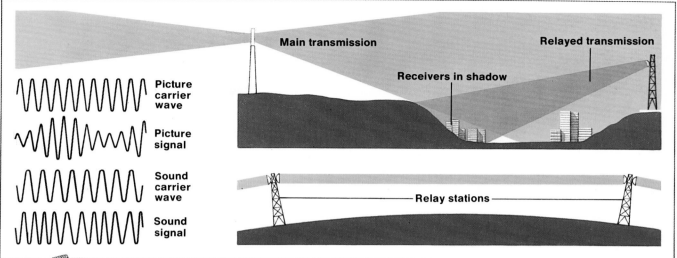

Picture carrier wave

Picture signal

Sound carrier wave

Sound signal

Main transmission

Receivers in shadow

Relayed transmission

Relay stations

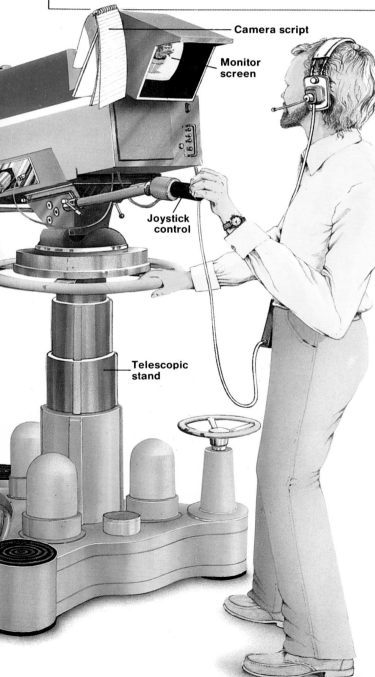

Camera script

Monitor screen

Joystick control

Telescopic stand

◀ The studio camera is big and heavy but can be moved easily. The cameraman can slide it backward, forward and sideways (tracking). He controls the lens from the joystick, zooming closer to the subject or pulling back, without moving.

▲ At the transmitter, the engineers use two oscillators to produce carrier waves. One wave is mixed with the picture signal, the other with the sound signal. These are strengthened (amplified) and mixed together before being sent out into the air.

made up from all the colors of the rainbow. In the color television camera, there is a series of filters called dichroic mirrors which split the light into beams of red, blue and green light. The beams are transformed into electrical signals.

Transmitting the Signals

The three tubes first mix the light to produce a signal called the luminance signal. This carries only light and shade and can also be received by black-and-white sets. Then the tubes produce the color or chrominance signal. Both signals are sent to the transmitter.

At the transmitter, the engineers create two radio waves called carrier waves. The sound signal which matches the picture is carried by one wave. The picture signals (luminance and chrominance) are carried by the other.

Finally, the two carrier waves themselves are combined and the signal is sent out. The carrier waves can't travel through mountains or buildings, and they need strengthening every 60 miles (97 kilometers). So they are sent across a network of transmitters and booster stations.

On the Studio Floor

The picture you see on television at home is only a tiny part of what's going on in the studio. Pull the camera back to "widen the shot" and you will see much more.

The Cameramen

First you will see four or five cameras worked by people wearing headphones. In spite of their size, television cameras can glide like swans and move at the touch of a fingertip. Sometimes a camera is mounted to swing in the air for spectacular shots.

Each cameraman has a small monitor on his camera to show him what's being transmitted to the viewers, as well as a viewfinder to see his own picture. He can change shot by moving the lens in and out (zooming) and by moving the camera along the floor (tracking).

The Director and Floor Manager

The director gives all the orders in the studio. He or she gives instructions through a microphone and uses special words to say what kind of shot is required. For example, if the director asks for a "BCU" (big close-up) of someone, he or she wants the cameraman to show the person's face only.

The person rushing around the studio wearing headphones is the floor manager who organizes everything taking place on the studio floor. He or she takes care of the program guests and generally makes sure that everything runs smoothly. The studio itself, especially if there is no audience, is divided into areas so that the studio staff can move from one item to another without stopping.

The director of a children's magazine program might divide the studio into three separate areas. The smallest may have a desk for a newsreader or frontman to introduce items. Another might have some screens, called "flats", to display material such as paintings, and the third area, a large space near the studio doors, might be used to stage big items.

Sound and Light

Sound is just as important as pictures. Well before the program begins, the sound people have been down on the floor fixing microphones, or "mikes", to the desk. If the mike swings overhead it is known as a boom.

Lights are positioned on the studio ceiling. These have shutters, or "barn doors", in front of them to control the amount of light needed. The electricians (called "sparks") adjust them with long poles. The entire set is controlled by a lighting panel so that different effects and colors can be created for different items.

Waiting well "out of shot" are the scenery people who move large items like the flats, and the property people whose job is to move small things such as the newsreader's chair.

A flashing red light in the corner of the studio shows that there's just one minute before the program goes on the air. It is time to move upstairs into the gallery, or control room. From now on, the director controls the studio through the floor manager, who receives instructions through headphones.

▲ Cameramen wear headphones so that they can receive instructions from the director in the control room.

▶ Shooting drama in the studio. Each camera gives a different view of the set. The microphones and lights are out of shot.

In the Gallery

In the final seconds before a program begins, the director may wish everyone "Good luck!" over the microphone. So many things can go wrong and everyone has to concentrate very hard all the time.

Who's Who in the Gallery

The director sits in front of a control panel called a mixing desk. On one side of the director is the vision mixer who "cuts up", or produces, the picture the director asks for by operating control keys and switches.

On the other side of the director is a production assistant who constantly tells everyone, including the frontman, how many seconds there are left on an item. The frontman listens through an earpiece invisible to the viewers. He or she can then read at the correct speed to fit in with the recorded material.

At least two other people sit at the control desk, although it isn't so obvious what they do. They are in fact the first and second engineers, who are in charge of everything – from lighting to the technical quality of pictures being sent to the gallery from outside the studio.

The Monitors

Above the control desk are lines of screens called monitors. These are not television sets, but closed circuit screens which show the

◀ The gallery, or control room. The director sits in front of the control desk, and uses a microphone to talk to the floor manager and cameramen on the studio floor. The bottom row of monitors shows the picture from each camera. The vision mixer cuts to each picture as the director asks for it.

▶ The control room of a television news program. Many of the items are broadcast live, so everyone has to think and act quickly. The newsreader in the studio receives instructions from the director through an earpiece which is not visible to the viewers.

director all the alternative pictures he or she can use. Only the central screen shows what is being transmitted.

The picture from each camera is shown on a separate monitor. One may show the program frontman, another the newsreader, while the third and fourth may show the empty demonstration area from different angles. A fifth camera may be all set up ready for the weather forecast.

Other monitors may be showing very strange things. A figure six on one comes from the film area to show that a film is loaded on the projector and is ready to go. This has to be started six seconds before the picture is ready, to reach the right speed, rather like an athlete running up to a jump.

Another monitor may show a clock. It is there as a symbol of the beginning of a video tape item. Like film, standard video needs a few seconds to get up to the right speed.

Another screen shows white lettering only. This comes from a minicomputer which produces words to identify people or things on the program. The words are superimposed on the pictures appearing on the screen. If slides are being used on the program, they are shown on the telejector machine.

The Sound Engineers

In a separate box near the control room are the sound engineers. Just as the director cuts from camera to camera, so the sound engineers control the microphones, add special sound effects and "fade" the sound in and out. It can be disastrous to leave a microphone on at the wrong time, as people will often say silly things if they think no one can hear them.

Outside Broadcasts

It is easier to move television out of the studio than it looks. The whole gallery can be packed away in a vehicle called the scanner and driven to wherever the program needs to be. A big outside broadcast (OB) unit looks like a traveling circus.

The Outside Broadcast Unit

The staff are much the same as in a normal studio. They include engineers, the director, and lighting people, plus a very important character called the rigger driver. Rigger drivers drive the scanner, and the generator vehicle if there is no electric power at the site. They may also drive other back-up vehicles, some with video machines that can record material and edit it on the spot.

The engineers have an extra responsibility outdoors. They have to find a suitable spot for an antenna to send the pictures back to the television studio, from where they are sent out around the country. Some vehicles carry a dish antenna. If the signal is blocked by buildings, the engineers take the dish off and fix it in a high position.

Sports and Instant News

When television producers first discovered the excitement of taking the machinery outdoors, they went everywhere creating television "firsts" – broadcasting from mountains, aircraft and ships. Nowadays, OBs are usually kept for events which can't take place in the studio, such as sports.

One of the most interesting effects of OBs is that history sometimes happens live on television. Presidents die, men land on the Moon, and records are broken. The outside broadcast is a wonderful help to the news reporter, and this kind of outside broadcast is increasing. New developments in video electronics have brought smaller and better equipment so that OB units can get better pictures back to the studio much more quickly.

▼ A mobile ENG crew. Their recorded material can be rushed to the studios, or it can be edited on the spot and transmitted from the portable antenna.

▶ An outside broadcast unit. The scanner vehicle houses the monitors and a mixing desk. The antenna sends the pictures back to the studio.

Electronic News Gathering

The introduction of electronic news gathering (ENG) is replacing film coverage, particularly in North America. For ENG, a portable video camera is used. Sometimes it is used like film: material is shot, taken back to the studios and edited. At other times it is edited on the spot, and is very often transmitted live into news bulletins.

ENG trucks carry a pop-up antenna to send the picture back, and editing machines for the video tape. Two cameras can be brought out of the truck on cables or they can wander free using battery power. This is called the ENG mode.

ENG is very useful during a difficult situation such as a riot. While the truck is parked safely in a side street, the cameraman and his sound recordist dash around taking video pictures. It takes only a few seconds to take the video tape out of the recorder, or "deck", and as soon as the crew get back to the truck they beam the pictures back to the television station.

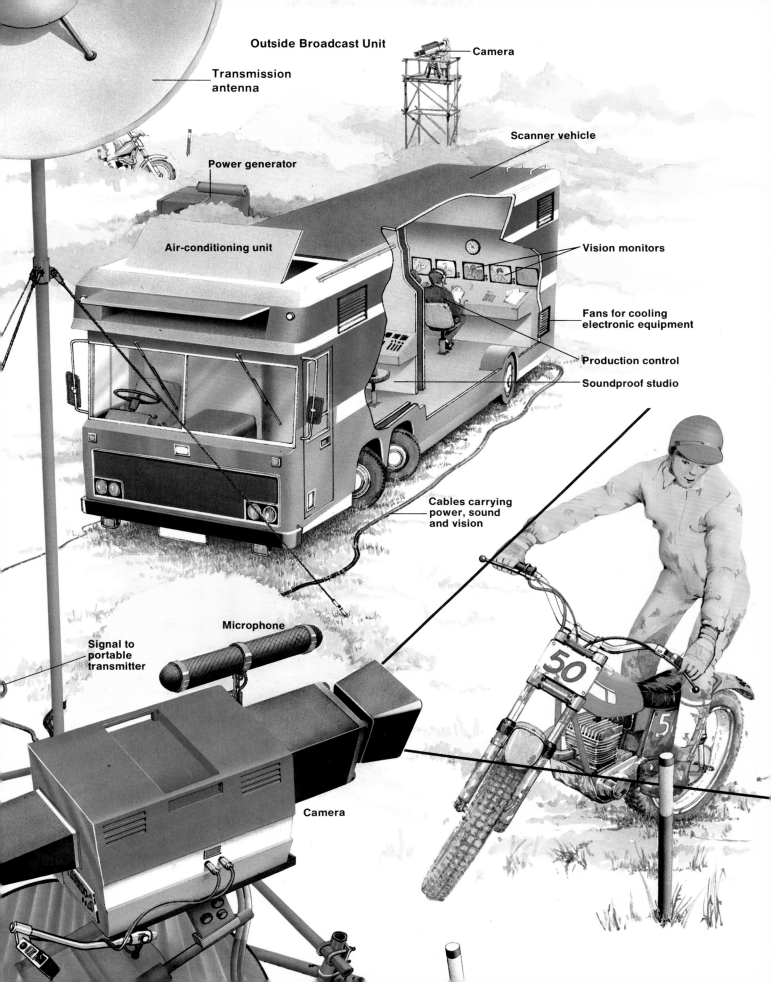

Outside Broadcast Unit

Camera

Transmission antenna

Scanner vehicle

Power generator

Air-conditioning unit

Vision monitors

Fans for cooling electronic equipment

Production control

Soundproof studio

Cables carrying power, sound and vision

Microphone

Signal to portable transmitter

Camera

Film and Video Tape

There is a well-known television story about the director whose program was going wrong. Pre-recorded film and video items did not appear. He kept completely calm in the chaos and finally leaned forward and said to everyone listening: "Okay, anyone who's got anything reeled up, run it!" He got the closing credits only four minutes after the program had begun!

Pre-recorded and Live Programs

Television is almost too good to be true, with perfect presentation and actors who don't forget their lines. It is, of course, so good because most of it is recorded, not "live", and the mistakes are edited out long before the programs are broadcast. Even live programs such as newscasts have many pre-recorded items.

Recording on Film

Film is the oldest recording medium used in television. The normal film crew includes a cameraman and assistant, a sound man (with an assistant if needed) and one or more people for lighting. They usually use 16-millimeter-wide film (twice as wide as home movie film) of two different kinds.

Ektachrome is the cheaper and easier film because only one developing process is needed, and it produces an image which can be broadcast directly. There is only one original

▶ A video editor never cuts into the tape. He or she re-records the wanted sections on to a new tape by using two machines called an edit pair. Making a series of smooth edits is a job for professional machinery. With home video equipment it is very difficult to make edits that don't make a noise or "jump".

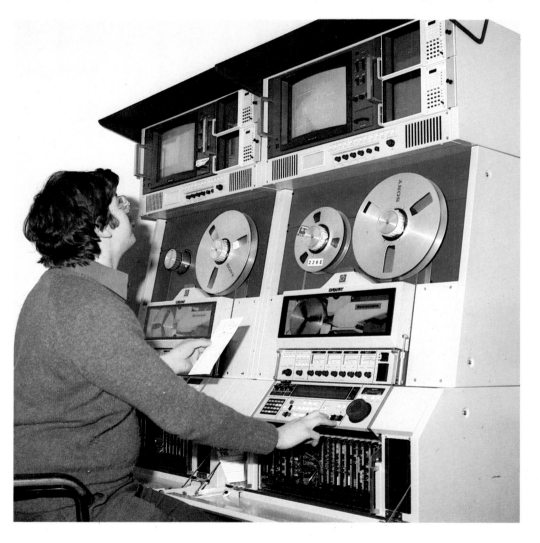

which is cut and transmitted. The disadvantage of Ektachrome is that the image can easily become scratched and dirty with use.

The other 16-millimeter film is called Eastman Color. This produces a negative image from which a "cutting" print is made. When the print has been altered and approved, the editor goes back to the original and gets a brand-new print. He or she matches it with the cut version by matching numbers on the edge of the film. Look closely when you are watching television and you may be able to spot the difference between Ektachrome used in news programs and Eastman Color used for longer films with beautiful scenery.

Storing Pictures on Tape

A more recent way of storing television pictures is electronically, on video tape. There are different sizes of tape which produce pictures of varying quality. Many engineers will say that the wider the video tape, the better the picture.

At the moment, television companies use a mixture of machines, some with two-inch (50-millimeter) tape, some with one-inch (25-millimeter) tape, and some with three-quarter-inch (20-millimeter) tape. Sound and pictures are on the same tape and the editing process is different from film.

Editing Film and Video Tape

The film editor works by using a machine that cuts the film into strips and sticks it back together again. Unwanted film is simply cut out. Video editing is done by re-recording sections in the right order on to another machine. The tape itself is never cut. When you have finished with the material on tape, it can be erased and the tape is ready to be used again.

Film, of course, is permanent and is therefore much more expensive than video tape. On the other hand, film cameras and editing machines are much tougher and last very much longer than video equipment. And there's the argument about the quality of the picture. Although the video picture is clearer, directors still prefer to use film to achieve soft, romantic effects.

▼ A film editor cuts the film and places the edges together in a joiner. The film is held in place by sprocket holes in the joiner which match the holes in the film. The editor puts special sticky tape over the edges of the film and punches the joiner down to stick the film together.

▲ The film editor also matches the pictures with the soundtrack. To help with this, a clapperboard picture appears at the beginning of each shot. As the board closes, the editor finds the matching noise on the soundtrack. All the pictures and sounds which follow will also match.

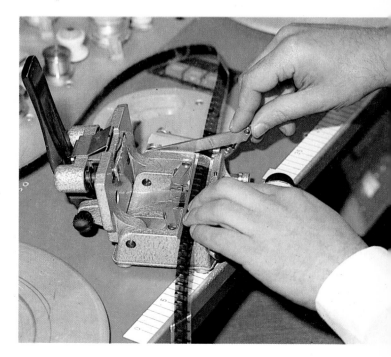

Satellites and Cable

The family of the future will have many more television channels than we have at present. There are comparatively few because national television channels are duplicated all around the country and this uses up the available frequencies. To have more channels, other methods must be used, such as signals transmitted by satellite or channelled along cables into the home.

Beaming Pictures by Satellite

Television people call satellites "birds". They use them all over the world, especially for news and sport. Satellite transmission is like passing a ball to your partner by bouncing it off the ceiling. The television station sends its picture to a ground station, which sends it up to a satellite. This bounces it back down to the ground station in whichever country the signal is going to. From here it passes along ground lines or it can even be bounced along more dishes (called microwave dishes) to the television station. Then it is sent out to the transmitters and the viewers.

International satellites are like aircraft with bookable flights. Most of the time available on them is booked by the big three US television networks: NBC, CBS and ABC. Quite often they book more time than they need, to try to stop their competitors. Smaller companies in the rest of the world can "borrow" time. Suppose an American network has a booking to get pictures from a sporting event in Turkey. European countries can get a "downleg" of these pictures to their own ground stations as they pass over on the way to the United States.

Direct broadcast satellites (DBS) work in a similar way, but produce a much stronger signal. To receive pictures from an international satellite you need a receiving dish bigger than a house. To get DBS pictures, a dish about four feet (one meter) across is big enough. You need permission to place it correctly in the yard to work efficiently, as well as an expensive alteration to your television set.

Cable Television

A big extension of cable television is also planned. Cable isn't a new idea. It is used throughout the United States. This kind of distribution is particularly useful in mountainous areas because the mountains interfere with airborne signals.

Copper coaxial cables carry television pictures and sound as electrical signals. These signals have to be boosted every few miles. Now there is a new kind of cable called optical-fiber cable. This is made of bundles of stretched-out glass which can carry signals uninterrupted for longer distances. The electrical signals are converted into pulses of light which travel along the thin glass fibers inside the cable. Each cable can carry many television channels.

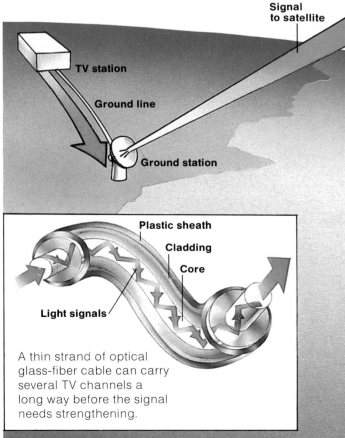

Signal to satellite

TV station

Ground line

Ground station

Plastic sheath

Cladding

Core

Light signals

A thin strand of optical glass-fiber cable can carry several TV channels a long way before the signal needs strengthening.

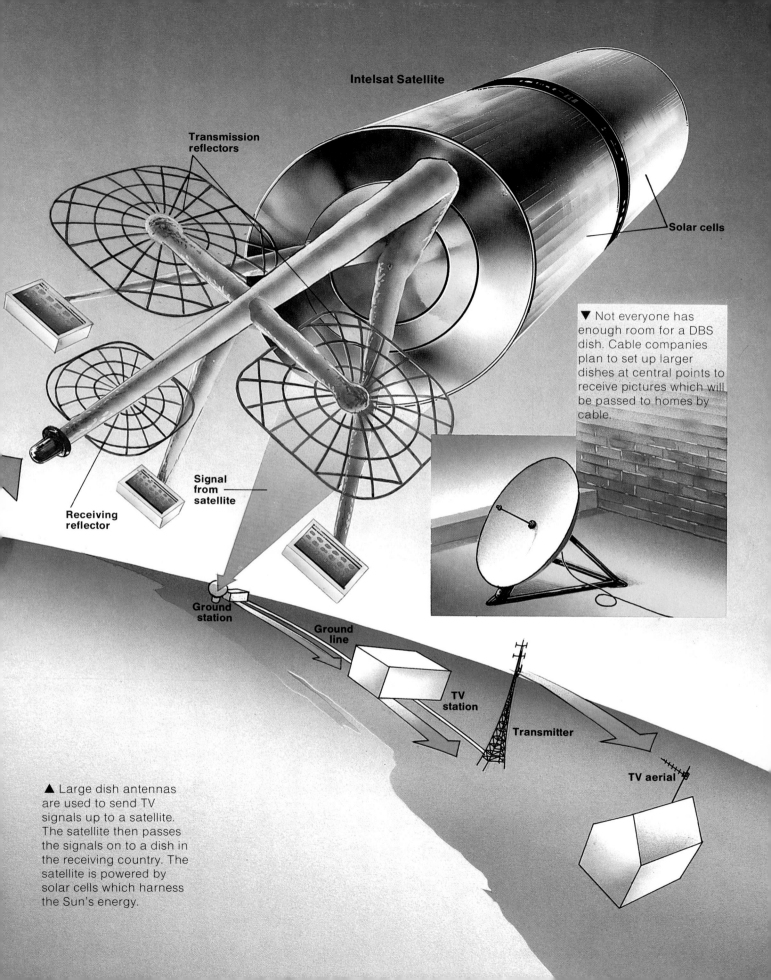

Intelsat Satellite

Transmission reflectors

Solar cells

Receiving reflector

Signal from satellite

Ground station

Ground line

TV station

Transmitter

TV aerial

▼ Not everyone has enough room for a DBS dish. Cable companies plan to set up larger dishes at central points to receive pictures which will be passed to homes by cable.

▲ Large dish antennas are used to send TV signals up to a satellite. The satellite then passes the signals on to a dish in the receiving country. The satellite is powered by solar cells which harness the Sun's energy.

Developments in TV

A television set is packed with complicated electronic equipment to decode the signals and show us a picture we understand. It is difficult to make smaller and thinner sets because of the size and shape of the cathode ray tube. To make flat screens, the tube has to be redesigned, or a different method has to be invented to decode the picture signal.

Making Screens with Lasers

A lot of time and money have been spent looking for a replacement for the cathode ray tube. A French team has made great progress by using a coating of silicon on glass. (Silicon is the material used for making microchips.) They use a ruby laser to turn areas of the silicon coating into crystals. Finally, using normal microchip methods, they turn the crystals into transistors, which are tiny electrical switches.

Wristwatch Television

Wristwatch television is also on its way. People will be able to buy miniature sets that show them breakfast television on the way to work. These sets will also tell you the time and date, and you may be able to play space invaders with them too. This may be a short-lived craze. For example, you can print books so small that you need a magnifying glass to read them. Tiny television could be just as frustrating.

▲ The eidophor system uses a strong light beam and slatted mirrors to project television pictures off a thin film of oil on to a large screen. Because of the high cost, this system is not practical for use at home but is used in places such as sports stadiums.

▶ In present 3-D TV, two identical images are seen together through special viewing spectacles (1). Another method will use a double television tube to project two images (one for each eye) on to a grooved lens (2). The grooves direct the right information to each eye.

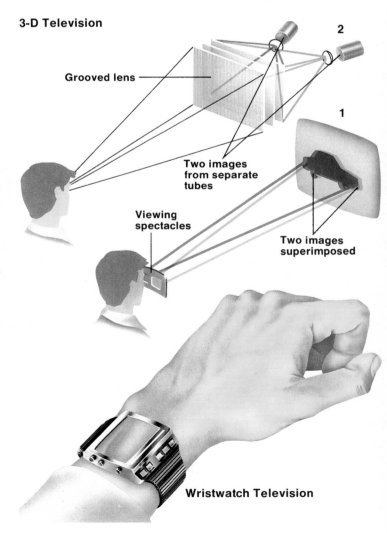

3-D Television

Grooved lens

Two images from separate tubes

Viewing spectacles

Two images superimposed

Wristwatch Television

3-D Television

Three-dimensional television works by pointing two cameras at the same object from different angles, then superimposing the two pictures. You can see depth, and objects seem to come out of the screen at you.

Three-dimensional photography was tried out many years ago in the cinema and has recently become popular again, now that new equipment has been developed. It is particularly in demand for horror films, when it is really exciting to imagine a vampire coming out of the screen towards you.

The major problem with 3-D photography is that the eyes need help to put the two separate images together, and at the moment viewers have to wear special glasses. The problem is worse for television than for the cinema with its big screen. With a small screen, you have to keep your head still or the picture breaks up.

The latest 3-D experiments are designed to get rid of the glasses that viewers have to wear. Instead, engineers are trying to use a double television tube with a special grooved lens in front to create an instant picture from the two.

The Effects of Television

The biggest future development is not really technical, but about how our lives may change because of television. Some doctors are already worried about children who spend too much time playing with home computers and video games all by themselves. If people can do a lot of work at home using television, and have no need to go out unless they want to, will they perhaps become people who meet only through television?

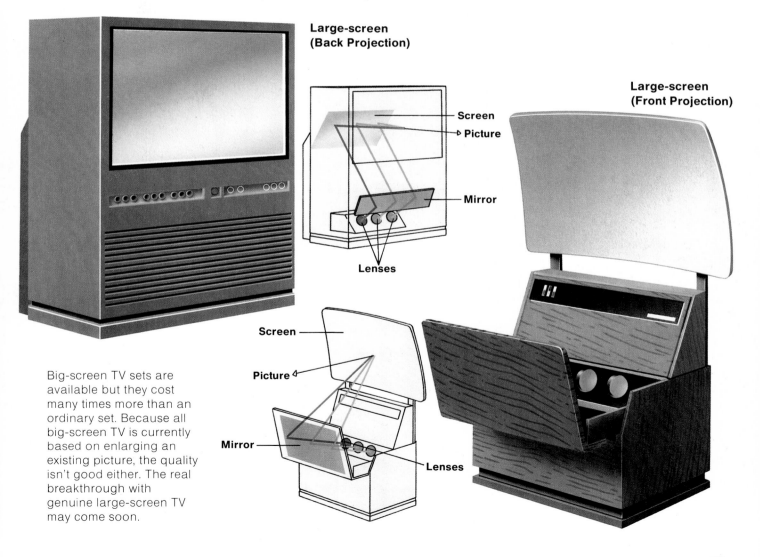

Large-screen (Back Projection)

Screen
Picture
Mirror
Lenses

Large-screen (Front Projection)

Screen
Picture
Mirror
Lenses

Big-screen TV sets are available but they cost many times more than an ordinary set. Because all big-screen TV is currently based on enlarging an existing picture, the quality isn't good either. The real breakthrough with genuine large-screen TV may come soon.

Video Recorders and Video Disks

The first home video machines appeared in the early 1970s. They used one-hour tapes and had simple controls. Now video machines can be programmed up to 14 days in advance, and can record on different channels.

Recording on Tape

There are several types of home video-tape systems. Unlike records or sound cassettes, however, you cannot switch video tapes from one system to another, even though the machines work the same way.

Picture recording is similar to sound recording. They both use magnetic tape. A recording head in the video recorder transfers an electronic signal to the tape by altering or disturbing the magnetic field on the tape. The signal is replayed by doing the same thing again, but in reverse.

As we have seen, a picture signal is more complicated than a sound signal and it needs more space. So video tape is wider than music tape. The head records the signal in a diagonal pattern across the tape, in strips close together. This enables you to get more on. You could record the picture in a straight line, but it would take 20 miles (33 kilometers) of tape to make a one-hour recording.

Video Disks

The video disk is different from video tape in several ways, the most important one being that you can't record on a disk at home. It is rather like the difference between records and cassette tapes for music. Disks have been used professionally for several years for sports "action replays", but are new to most people.

The leading system of picture recording on disk uses lasers. The disk is made of plastic and is printed with tiny pits which hold the picture and sound information. Instead of a needle touching the disk, a laser beam is used to read the information in the pits. The beam is constantly directed towards the pits by a series

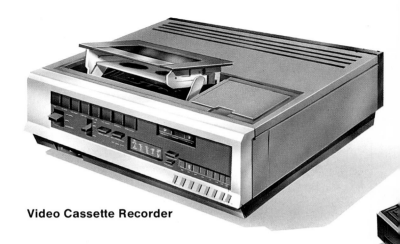

Video Cassette Recorder

of mirrors and prisms. Most people agree that the picture quality of disks is better than that of video tape.

Making Disks and Tapes

Disks should be quicker and cheaper to make than pre-recorded tapes because each disk is stamped out with a complete program on it, rather like an LP record. Video tapes are produced by copying another video tape. This process takes as long as the program lasts, and cannot be speeded up. Sometimes, because of careful checking at each stage of production, disks may also take a long time to make too, so it is difficult to say which system has the most advantages.

Video-disk Player

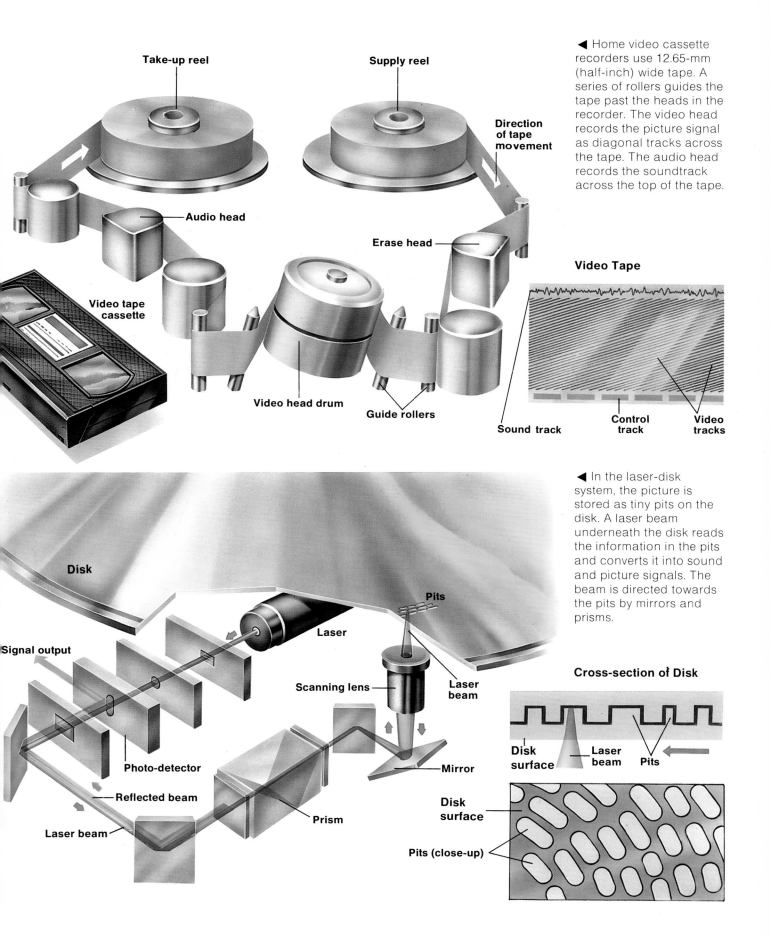

Take-up reel

Supply reel

Direction of tape movement

◄ Home video cassette recorders use 12.65-mm (half-inch) wide tape. A series of rollers guides the tape past the heads in the recorder. The video head records the picture signal as diagonal tracks across the tape. The audio head records the soundtrack across the top of the tape.

Audio head

Erase head

Video Tape

Video tape cassette

Video head drum

Guide rollers

Sound track

Control track

Video tracks

◄ In the laser-disk system, the picture is stored as tiny pits on the disk. A laser beam underneath the disk reads the information in the pits and converts it into sound and picture signals. The beam is directed towards the pits by mirrors and prisms.

Disk

Pits

Signal output

Laser

Cross-section of Disk

Scanning lens

Laser beam

Photo-detector

Disk surface

Laser beam

Pits

Reflected beam

Mirror

Prism

Disk surface

Laser beam

Pits (close-up)

Making a Video: Equipment

Most home video recording machines can be fitted with a camera that can be plugged into it through a camera adapter box. This system isn't portable, so you may have to stay in the house or use your camera in the yard with the help of an extension lead. Portable cameras and recording decks to match are available, but they cost more than home movie cameras.

The Advantage of Video

There is a great advantage to using video equipment instead of film. If you make mistakes with film, you don't find out what's wrong until it has been processed, and that may mean you have wasted a lot of time and money. With video, you can experiment with shots to see how they look as they are being taken. You check the shots by replaying them on the recorder. If there's anything wrong, you can erase the shots on the tape and start again. The same tape can be used over and over again until it's right.

The Camera

There are two important things to remember. Read your camera instructions carefully and, if you can't understand any of them, get expert advice. Never take things to pieces to "make them work". Second, a video camera is not very strong compared with a still camera or home movie camera. It has a delicate tube like those in television studio cameras, and you can break the tube by dropping or bumping the camera. You can "burn" the tube too. Burning happens when the iris, or aperture, is left open and the camera is accidentally pointed at the sun or at a strong light.

Microphones

Sound varies from camera to camera. If possible, always plug in a separate microphone if your system allows it. The "mike" built into the camera is a very poor method of collecting sounds.

Advance Planning

Making a video is not really something you can do by yourself. You need several friends to help you. Give everybody a job before you start. You will need actors, a camera operator to shoot the pictures, and a producer to write the script and see that everybody is ready. The sound recordist's job is to hold the microphone in the right position. The properties and continuity person checks all the details. The engineer watches the television screen and makes sure that the equipment is working perfectly.

Work to a definite plan because you can only pause between shots once you start the camera. The picture will blur every time you switch off and on again.

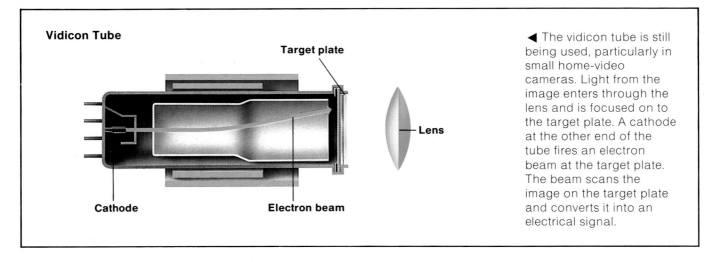

Vidicon Tube

Target plate

Lens

Cathode

Electron beam

◄ The vidicon tube is still being used, particularly in small home-video cameras. Light from the image enters through the lens and is focused on to the target plate. A cathode at the other end of the tube fires an electron beam at the target plate. The beam scans the image on the target plate and converts it into an electrical signal.

Video cameras are simple to operate. There are three main controls to learn about. The iris, or aperture, controls the amount of light going into the camera. This control may be automatic or manual. The focus control is used to make sure that each shot is sharp. It is usually manual. The zoom is a way of closing in to or pulling out from the subject without losing the sharpness, or "focus". The zoom control may be manual or automatic.

Microphone

Lens

Focus ring

Zoom lever

Start/stop

Aperture control

White balance control

Viewfinder

Pistol grip

Wrist strap

Another important control is the "white balance". This is used to adapt the camera to different lighting conditions. On cheaper cameras, this control will be manual. Always check each shot on your TV screen and adjust the control according to the maker's instructions.

Making a Video: Shooting

The beginner should start making videos outside in the yard on a sunny day. Connect the camera to the power points inside the house with an extension lead. Always ask an adult to check that the plugs and connections are safe.

Using the Microphone

The sound recordist's microphone should be connected to the video recorder inside the house via an extension lead. The best way to shoot drama is to tie the microphone to a stick. The sound recordist can then hold it over the heads of the actors, taking care that the mike is low enough to record the sound but doesn't get into the picture. The continuity person should check that the actors know their lines.

Using the Camera

The camera operator should try to sit opposite the actors, or lean against a wall to steady the camera. The less jerky the camera movement is,

the more professional-looking your video will be. Make sure the sun isn't behind your subjects or the picture will come out looking black.

The camera operator should also practice different shots such as holding, panning, zooming and tilting. For the hold, steadily point the camera at the subject without moving the camera or the zoom control. To pan, move the camera horizontally across the subject from left to right or right to left. The tilt is similar to the pan, but is a vertical movement. To zoom, you usually press the zoom button on the camera. This closes in to a subject, or pulls out from it. Zooming should be used only when necessary. Even among professionals there is a tendency to zoom too much.

Because the picture wobbles for the first few seconds after the camera has been switched on, you should shoot your sequences in the correct order. Use the pause button to alter a shot when you cannot pan or zoom to change shot.

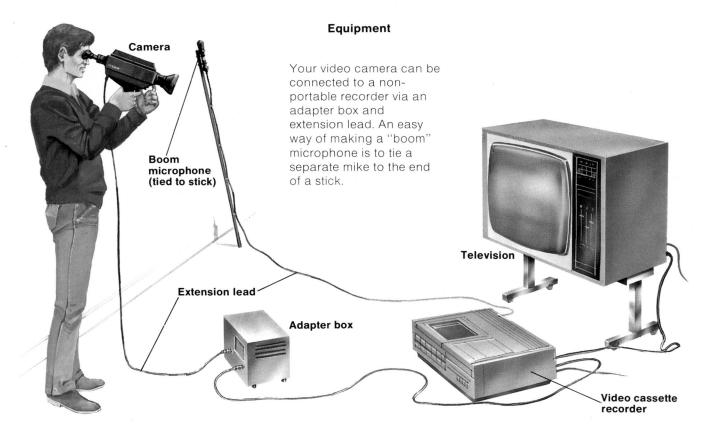

Equipment

Your video camera can be connected to a non-portable recorder via an adapter box and extension lead. An easy way of making a "boom" microphone is to tie a separate mike to the end of a stick.

Camera

Boom microphone (tied to stick)

Television

Extension lead

Adapter box

Video cassette recorder

Camera Shots

Long shot

▲ To hold, point the camera at the subject steadily. A long shot shows background detail.

Mid-shot

▲ A mid-shot shows more of the subject and cuts out some of the background.

Close-up

▲ A close-up cuts out the background and concentrates on the subject.

Panning

(1)

(2)

(3)

Tilting

(3)

(2)

(1)

▲ To pan, steadily move the camera horizontally across the subject. Do not make the panning movement too quickly.

Zooming

▶ The zoom enables you to close in to or pull out from the subject without moving the camera. The zoom should be used only when necessary.

◀ The tilt is similar to the pan, but is a vertical movement up or down, and is good for tall subjects.

Video at Work

The most common use for video equipment at work today is security. We have become quite used to being watched by cameras in stores. The old-fashioned store, where a sales assistant stood behind a counter and handed the customer each item, didn't need a camera "spy". But modern supermarkets are designed to tempt us into picking things up and paying for them later. Unfortunately, this means that some people are tempted into picking things up and not paying for them at all!

Security and the Law

Security cameras are small, simple and usually give black and white pictures. They are connected to a monitor, not a television receiver, which can be watched by the shop security staff. Camera systems are also widely used in banks.

Video technology has now been introduced into the courts, especially in North America. Cameras are good "detectives", and the evidence they provide shows the criminal

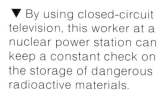

► The police now use video to help them with traffic control. They can see from a glance at the screens how traffic is moving, and can take immediate action if necessary.

▼ By using closed-circuit television, this worker at a nuclear power station can keep a constant check on the storage of dangerous radioactive materials.

actually at work on the crime.

Video cameras can also protect suspects who have been arrested. An interview at the police station can be video-taped with a clock in the background to prevent the tapes being edited or altered in any way.

Domestic security also can be improved with video cameras. Some local councils have installed them at the entrances of apartment blocks to cut down vandalism and attacks on old people. These cameras are connected to a separate cable channel on television sets in the block, so that the residents can see who's coming and going.

Teaching and Training
Teaching by video will become as important as security. Operations, for example, can be watched on a screen by many medical students at once. Doctors would also find video records a great help instead of having only written notes about their patients. Sick people in hospital, and babies in incubators, can be watched on a screen when the nurse isn't actually in the ward.

Business companies and other organizations can record their work and use the tapes to train new staff. Sports trainers and dance teachers already find video very useful. Movements can

be recorded in close-up detail and then played back slowly to see how the performance can be improved. Sports teams can use video to re-live a game and look at all the mistakes they made. If a complaint is made after a horse race, video can be used to see whether a jockey has behaved properly.

Estate agents have started using video tapes to sell property. Instead of looking at pictures and then going to see a lot of houses, people can view a video tape to give them a better idea of what the houses would be like to live in. They can "visit" many houses in half an hour on video

before choosing the few they are serious about visiting.

The Future
Perhaps salesmen of the future who come to the door will invite you to see a tape, instead of carrying heavy samples of the goods. Catalogs may be transformed from heavy volumes into television programs, either replayed by special services on television or given to you to look at in the store.

With all these possibilities, could we become a society where everybody is "watched", whether they like it or not? It is a worrying question to which there is no simple answer. We will have to make sure that the reasons for watching us are very good ones.

◀ Dancers in training or rehearsing new roles can be recorded on video. By playing back the video later, they can study their performance and improve on it.

▼ Many organizations now use video for training new recruits. By recording lectures and demonstrations on tape, they can be replayed many times over to different groups.

SOUND SYSTEMS

The first sound recording was made in 1877. Since then, engineers have continuously improved the quality of sound reproduction. This introduction to hi-fi shows how all the parts of a sound system work. It also looks at radio broadcasting and communications, and how disks and cassettes are made.

How Hi-fi Systems Work

Recording at Home

Making Records

Radio Broadcasting

Hi-fi and Stereophonic Sound

In 1877, Thomas Alva Edison showed that it was possible to record and play back sound. Ever since then, engineers have been seeking ways to improve the quality of the replay. Hi-fi, or high fidelity, equipment can now recreate sound to a very high degree of accuracy.

Stereophonic Sound

The original sound systems produced sound on one channel and this came from a single loudspeaker. As a result, the system was known as monophonic, or "mono" for short. This was followed by the stereophonic system which carries the sound on two channels and reproduces it through two loudspeakers. One speaker radiates one group of instruments or voices, and the other another group. The different sounds therefore reach your ears from different directions, as they would if you were listening "live". Because of this the performance sounds more natural. More recently, an Ambisonics UHJ system has appeared. This allows you to control the direction of the sound right up to 360°.

What Makes a Sound System?

Sound systems for the home used to be simple. The equipment was packed into one large cabinet called a radiogram.

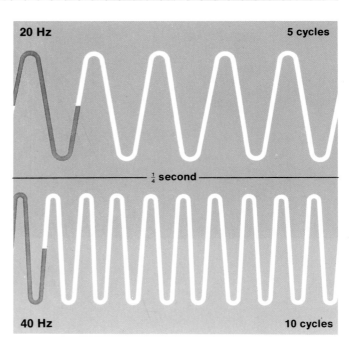

20 Hz ... 5 cycles

¼ second

40 Hz ... 10 cycles

▲ The frequency of sound waves is measured in Hertz (Hz). A frequency of one Hz equals one cycle (complete wave) per second. In the top diagram, 5 cycles occur in ¼ second, so the frequency is 20 Hz. A higher frequency of 40 Hz is shown below it.

▼ The main elements of a stereo system are shown below. Sound is carried on two separate channels and reproduced through two loudspeakers. Stereo attempts to give the listener the direction and spread of the original sounds. Stereo means "solid", or "three-dimensional".

◄ Hi-fi racks or towers are "separates" housed in a tall cabinet.

▼ Music centers have an amplifier, tuner, cassette deck and record deck in one cabinet.

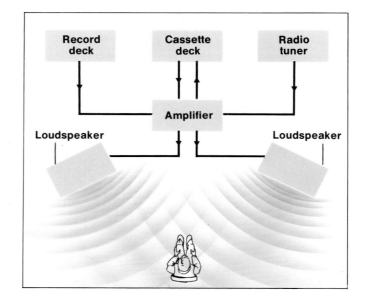

Record deck

Cassette deck

Radio tuner

Amplifier

Loudspeaker

Loudspeaker

Stereophonic Sound System

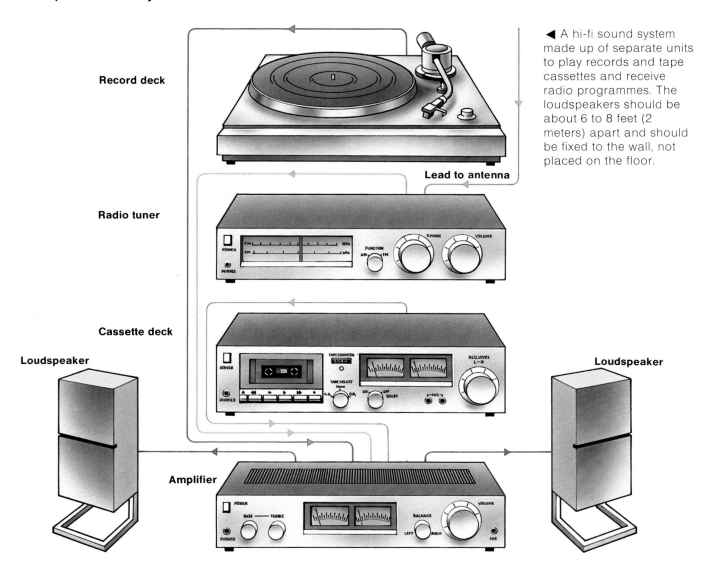

Record deck

Radio tuner

Cassette deck

Loudspeaker

Amplifier

Lead to antenna

Loudspeaker

◀ A hi-fi sound system made up of separate units to play records and tape cassettes and receive radio programmes. The loudspeakers should be about 6 to 8 feet (2 meters) apart and should be fixed to the wall, not placed on the floor.

Modern sound systems often have three groups of units, or separates. These can be connected to each other in different ways. This gives flexibility. You can start with a basic system and then gradually "upgrade" it by adding pieces to it and replacing existing units with improved versions.

At the "front end" of the system (the program source), there is a record deck for playing disks, a tape deck for recording and playing back sound, and a radio tuner. These send electrical signals to the amplifier (the heart of the system) which enlarges, or amplifies, them to drive the headphones and loudspeakers at the "back end".

Music Centers and Rack Systems

Hi-fi rigs with well-matched components can give superb sound quality. Several cables are needed to connect the units and to supply the power, so the back of a hi-fi system can easily look like a plate of spaghetti. To get around this problem, the music center was developed. It consists of an amplifier, radio tuner, cassette deck and record player linked together in one container. The only leads required are to the electricity supply and to the loudspeakers.

Hi-fi racks, or towers, house the units in a tall vertical cabinet. Each unit plugs to the one underneath. Mini, or micro, hi-fi systems are another trend. These are neat and compact.

Record Decks

Record decks all operate in much the same way. They consist of a turntable which spins the record, and a pickup arm with a cartridge which takes instructions from the disk and converts them to electrical signals.

The Turntable

The turntable has a circular platter which is rotated at a pre-set speed by an electric motor. For LP (long-play) disks and some 7-inch maxi-singles it has to revolve at $33\frac{1}{3}$ r.p.m. (revolutions per minute). For most 7-inch records it is slightly faster, at 45 r.p.m.

The drive system is the most important part of the turntable. Unless it operates silently without any changes in speed, it will affect the sound quality. A noisy motor, for example, will produce a distracting "rumble".

In one turntable model, a high-speed electric motor drives the platter with a rubber belt which passes around pulleys. More recently, direct-drive turntables have been introduced. In these the motor shaft is connected directly to the platter.

The Pickup Arm

The function of the pickup arm is to hold the cartridge in position as it travels across the disk. At the same time it has to apply correct tracking (downward) force. The amount of force will depend on the type and quality of the pickup arm itself. It can be adjusted by a counterbalance set at one end of the arm. The arm also has to be designed to minimize "bias" – a natural tendency to move toward the center of the spinning disk.

Pickup arms are usually pivoted and have an "S" or "J" shape. But this type has a small degree of "tracking error" because it does not follow the straight path across the disk that the cutter head followed when the recording was being made. However, you can now get decks with linear-tracking arms that move across the disk in a straight line.

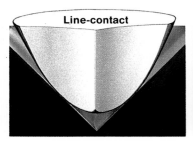

▶ The stylus vibrates as it traces the groove in a record. These vibrations are passed to the cartridge. The basic shape of styli is hemispherical, but elliptical and line-contact types are now popular.

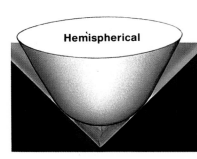

Hemispherical

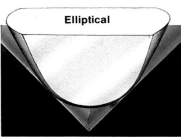

Line-contact

Elliptical

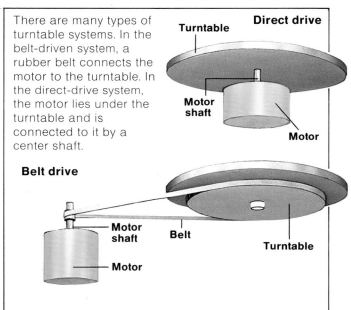

There are many types of turntable systems. In the belt-driven system, a rubber belt connects the motor to the turntable. In the direct-drive system, the motor lies under the turntable and is connected to it by a center shaft.

Direct drive

Turntable

Motor shaft

Motor

Belt drive

Motor shaft

Belt

Motor

Turntable

The Cartridge

Early radiograms and record players were fitted with crystal or ceramic cartridges. These have been replaced by the moving-magnet (mm), induced-magnet/moving-iron and moving-coil (mc) types. All these signal generators depend on the relative movement between a magnet and length, or coil, of wire. The "mm" type has a

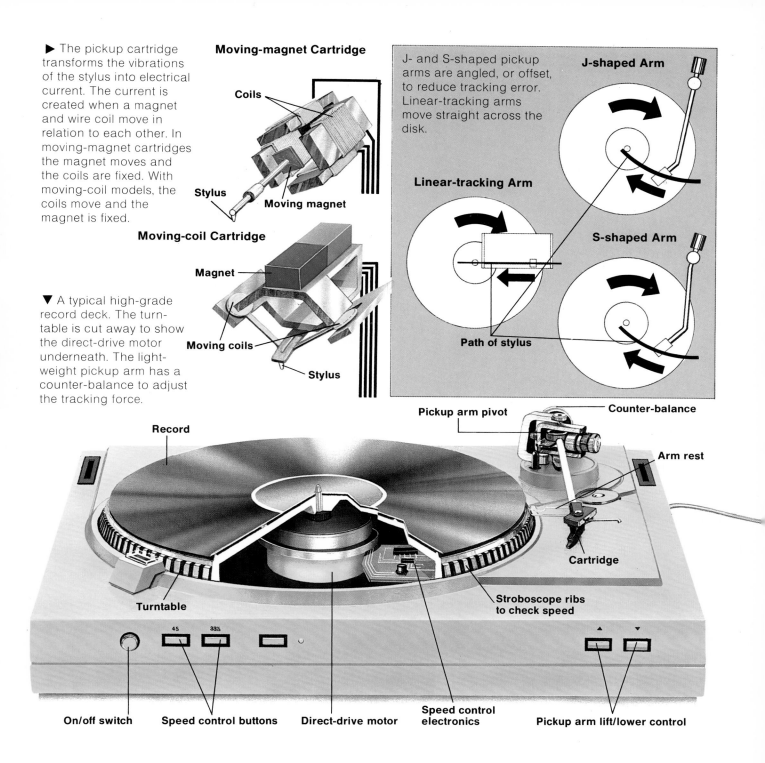

▶ The pickup cartridge transforms the vibrations of the stylus into electrical current. The current is created when a magnet and wire coil move in relation to each other. In moving-magnet cartridges the magnet moves and the coils are fixed. With moving-coil models, the coils move and the magnet is fixed.

Moving-magnet Cartridge

Coils

Stylus

Moving magnet

Moving-coil Cartridge

Magnet

Moving coils

Stylus

J- and S-shaped pickup arms are angled, or offset, to reduce tracking error. Linear-tracking arms move straight across the disk.

J-shaped Arm

Linear-tracking Arm

S-shaped Arm

Path of stylus

▼ A typical high-grade record deck. The turntable is cut away to show the direct-drive motor underneath. The lightweight pickup arm has a counter-balance to adjust the tracking force.

Pickup arm pivot

Counter-balance

Arm rest

Record

Cartridge

Turntable

Stroboscope ribs to check speed

On/off switch

Speed control buttons

Direct-drive motor

Speed control electronics

Pickup arm lift/lower control

fixed pair of coils. A tiny magnet mounted in the end of the stylus bar moves in response to the stylus vibrations and so creates a fluctuating current in the coils. In the "mc" type the magnet is fixed and the coils are set in the end of the moving stylus bar. Apart from these cartridges there is also a more recent, and more expensive, electrostatic design.

The Stylus

The playback stylus (needle) is attached to the cartridge. Its tip is made of sapphire or diamond, and is hemispherical in shape. But nowadays the elliptical (bi-radial) stylus is becoming popular. The large width of the tip prevents it from touching the bottom of the groove but keeps it in contact with the walls.

Cassette Decks

One of the biggest developments in hi-fi systems has been the emergence of the tape cassette. This was first introduced by Philips as the Compact Cassette in 1963.

The cassette deck uses a magnetic tape mounted in a plastic case to store the sound signals. When making a recording, the tape is drawn across a fixed recording head/amplifier. The signals put on the tape can later be "read" by one or more playback heads.

The cassette deck needs a drive system to wind the tape from one spool to the other, and it also has a number of features to improve the sound quality and avoid distortion. An essential feature is a recording-level meter. This can either be a VU (volume unit) or a PPM (peak program meter).

Every deck contains a bias oscillator which produces a high-frequency bias signal that you will not be able to hear on the tape. This signal is combined with the sound signal before it reaches the recording head. Its purpose is to help in reducing distortion in the recording system. It can also be used to wipe out (erase) signals just before the recording.

The Tape

The forerunner of the cassette deck was the open-reel or reel-to-reel tape recorder which had its origins in the Magnetophon developed in Germany in the late 1930s. The first tapes used were made of paper with a coating of fine magnetic particles. Later, plastic film bases were used.

Early cassette tapes had a ferric-oxide coating. In 1970, the finer-particled chromium-dioxide coating was introduced, followed by Ferrochrome. This consists of a two-layer coating of chrome on top of ferric particles. Metal alloy tapes are now available, but these function properly only on machines with suitable heads and electronics.

As everything is smaller and slower on a tape cassette, the tape makes a hissing noise. How-

ever, many decks are now fitted with a Dolby noise-reduction system. This was first developed for professional equipment. It works by compressing the dynamics (intensity changes) in signals before recording, and expanding them again on replay. The system was modified into Dolby B and Dolby C forms for cassette decks, and greatly improves the sound quality. Other systems that reduce noise from the tape include dBx, ANRS, ADRES and High-Com.

A system of C-numbering shows the running time in minutes of a cassette. For most purposes the C.90 is good value, giving 90 minutes (45 minutes on each side). Furthermore, the tape is not too thin and fragile. Other tape sizes are C.45, C.60 and C.120.

Micro-cassettes are being used more and more. These measure only 2 by $1\frac{1}{4}$ inches (50 by 30 millimeters). They were designed for note-taking and interviews, but their quality can be good enough for recording music.

Microphone socket
Headphone socket

Input-level control L/R
Bias switches
Equalisation switch

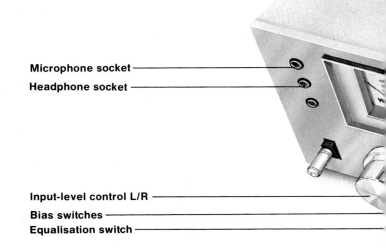

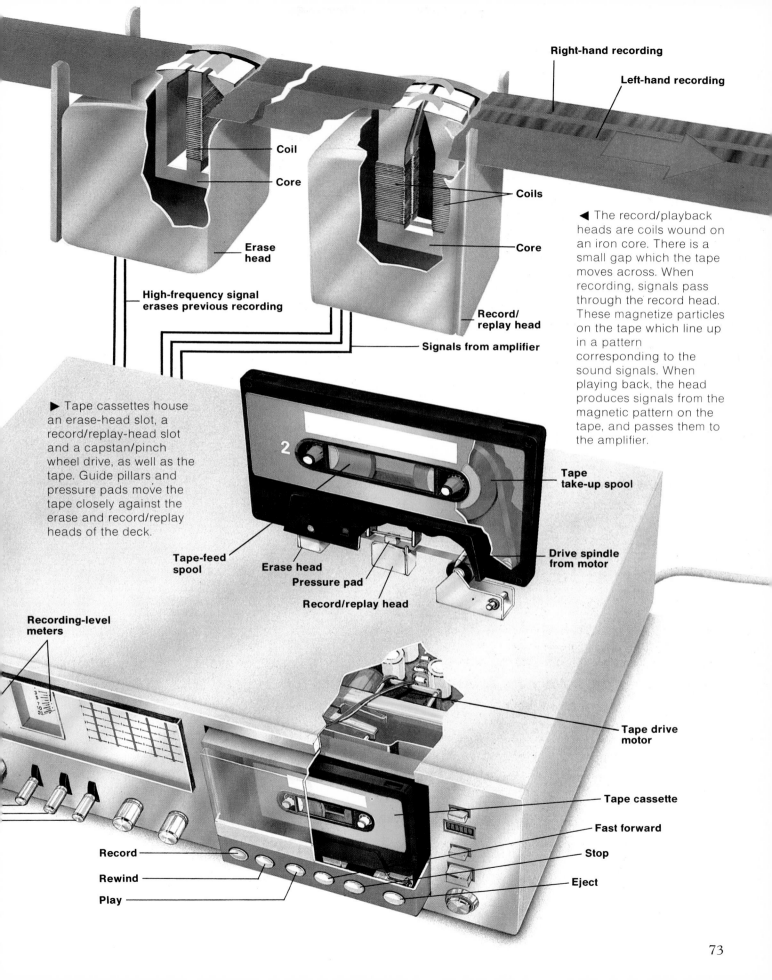

Right-hand recording

Left-hand recording

Coil

Core

Coils

Core

Erase head

High-frequency signal erases previous recording

Record/ replay head

Signals from amplifier

◀ The record/playback heads are coils wound on an iron core. There is a small gap which the tape moves across. When recording, signals pass through the record head. These magnetize particles on the tape which line up in a pattern corresponding to the sound signals. When playing back, the head produces signals from the magnetic pattern on the tape, and passes them to the amplifier.

▶ Tape cassettes house an erase-head slot, a record/replay-head slot and a capstan/pinch wheel drive, as well as the tape. Guide pillars and pressure pads move the tape closely against the erase and record/replay heads of the deck.

Tape take-up spool

Tape-feed spool

Erase head

Pressure pad

Record/replay head

Drive spindle from motor

Recording-level meters

Tape drive motor

Tape cassette

Fast forward

Record

Rewind

Play

Stop

Eject

Radio Tuners

The job of a radio tuner is to receive radio signals from transmitting stations. Tuners come in a variety of shapes, sizes and prices. Their performance varies too. The tuner may come as a separate unit, or it may be combined with an amplifier as a tuner-amplifier, or "receiver".

Transmission and Reception

A radio signal must carry two pieces of information about the sounds to be recreated at the listener's end. These are frequency and amplitude.

The original sound-wave form cannot be transmitted as a radio wave because the frequency is too long. So a carrier wave at very high frequency is sent out from the transmitter with the sound signal superimposed on it.

The two ways of impressing the sound signal on the carrier wave are AM (amplitude modulation) and FM (frequency modulation). In AM, the carrier *frequency* is constant and the carrier *amplitude* is varied. In FM, the carrier *amplitude* is constant but the carrier frequency is varied.

The antenna of your tuner picks up the carrier wave from the transmitter. This sets up weak electric currents in the antenna which vary with the carrier wave. The tuner separates the carrier wave from the sound signal and sends the sound signal to the amplifier.

Antennas

For FM and VHF (very high frequency) reception, it is best to use a multi-element antenna, particularly for stereo signals. If the signal is strong enough, you can use an indoor antenna or even the built-in rod on the tuner, which picks up AM signals.

Features of the Tuner

Many tuners now have a quartz-locked synthesizer to tune the radio stations accurately. This also keeps them locked in to make sure the signal

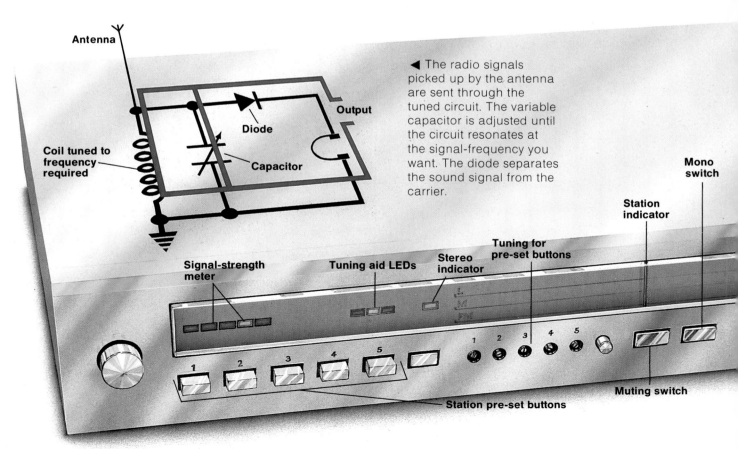

◀ The radio signals picked up by the antenna are sent through the tuned circuit. The variable capacitor is adjusted until the circuit resonates at the signal-frequency you want. The diode separates the sound signal from the carrier.

Antenna

Coil tuned to frequency required

Diode

Capacitor

Output

Signal-strength meter

Tuning aid LEDs

Stereo indicator

Tuning for pre-set buttons

Station indicator

Mono switch

Station pre-set buttons

Muting switch

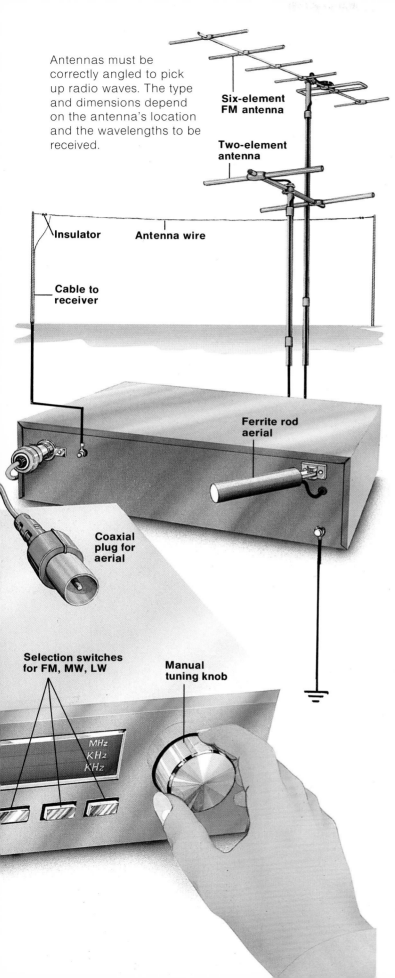

Antennas must be correctly angled to pick up radio waves. The type and dimensions depend on the antenna's location and the wavelengths to be received.

Six-element FM antenna

Two-element antenna

Insulator

Antenna wire

Cable to receiver

Ferrite rod aerial

Coaxial plug for aerial

Selection switches for FM, MW, LW

Manual tuning knob

MHz
KHz
KHz

does not drift. Advanced electronic tuners also enable you to pre-select stations. You program the radio station frequencies you normally listen to into the tuner's memory circuit. These can then be found at any time at the push of a button.

Manual tuning to an AM station is usually quite straightforward. You merely twist the tuning knob to the station you want and listen for the correct tonal balance. Tuning by hand to an FM station is more tricky and requires a little more fiddling to get the best noise-free signal.

To help the user, tuning indicators are often fitted. One gives the relative indication of FM and AM tuned signal strength (usually by a center zero point). Another shows the maximum signal strength when the station is precisely "on tune", that is, when the signal is at peak level. In some of the latest tuners, rows of LEDs (light-emitting diodes), or the tuning pointer on the dial, light up when the station is spot on frequency. These devices replace the meters.

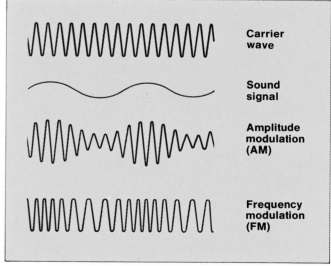

Carrier wave

Sound signal

Amplitude modulation (AM)

Frequency modulation (FM)

◀ Most features on radio tuners are connected with tuning. This can be manual, by turning a knob, or automatic. There may be pre-set buttons for changing stations quickly. Accurate tuning is helped by signal-strength meters, or LEDs.

▲ In AM, the amplitude (or size) of the carrier wave varies in relation to the sound signal. In FM, the amplitude remains constant, but the frequency varies with the sound signal. FM broadcasting gives better sound quality.

Amplification and Control

The amplifier is the heart of a sound system. Its basic job is to boost the signals from a disk record, tape or radio tuner, so that they are strong enough to drive the loudspeakers. The first feeble signal may be only one-thousandth of a watt, and it needs amplifying to 10, 20 or more watts. The amplifier is used to select the input into the sound system (from pickup, cassette deck or radio tuner) and also to provide controls for volume, tone and balance.

In some hi-fi systems the main amplification is done by a power-amplifier connected to a pre-amplifier or control unit. However, in many popular systems an integrated unit, which combines both these functions, is used. In another combination, a "receiver" houses a radio tuner, pre-amplifier and a power-amplifier.

Tone Controls and Fillers

Bass and treble tone controls are fitted on sound equipment so that the balance of frequencies reproduced in a room can be adjusted. This means that "tizzy" high notes or "boomy" bass sounds can be reduced.

In more advanced systems, the amplifier may be fitted with "filters". These can be used to remove low-pitched rumbles from the turntable, for example. High-frequency filters help in reducing unwanted whistles or station interference on radio signals, as well as surface-noise or distortion when playing disks or tape cassettes.

Connecting to the Amplifier

There is a wide variety of plugs and sockets for connecting the different units of a modern audio system. The most commonly used are shown on the opposite page.

Loudspeaker connections may consist of screw terminals or spring-loaded jaws to grip the bared wire of the loudspeaker cable. Alternatively, they may be two-pin DIN sockets, as fitted to European and Japanese equipment. Another popular method is a two-hole

socket, which accepts 4-millimeter phono plugs.

Jack plugs are usually used for headphone output and microphone input. Coaxial plugs are used for connecting radio tuners to antennas. Couplings between units are generally through DIN-type plugs and sockets. Make sure that the wire connections to a five-pin DIN plug are correct, to avoid switching the left and right channel sounds on a stereo system.

DIN refers to *Deutsche Industrie Normen*, a set of standards established in Germany, now adopted on equipment throughout Europe. On some equipment, manufacturers fit DIN and phono sockets wired in parallel to cope with all units and leads.

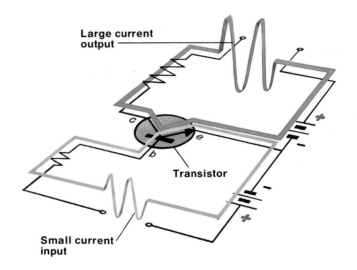

Large current output

Transistor

Small current input

▲ Amplifiers use transistors to boost, or amplify, signals. Three areas of a transistor have distinct electrical features. These are known as the base (b), emitter (e) and collector (c). A small varying current flowing between the base and emitter causes large variations in current between the emitter and collector.

▶ Few amplifier controls are really essential. Some models simply have an on/off switch and volume control. The balance control is used to adjust the sound level from each speaker. Tone controls correct the frequency balance between the recording and the room's acoustics.

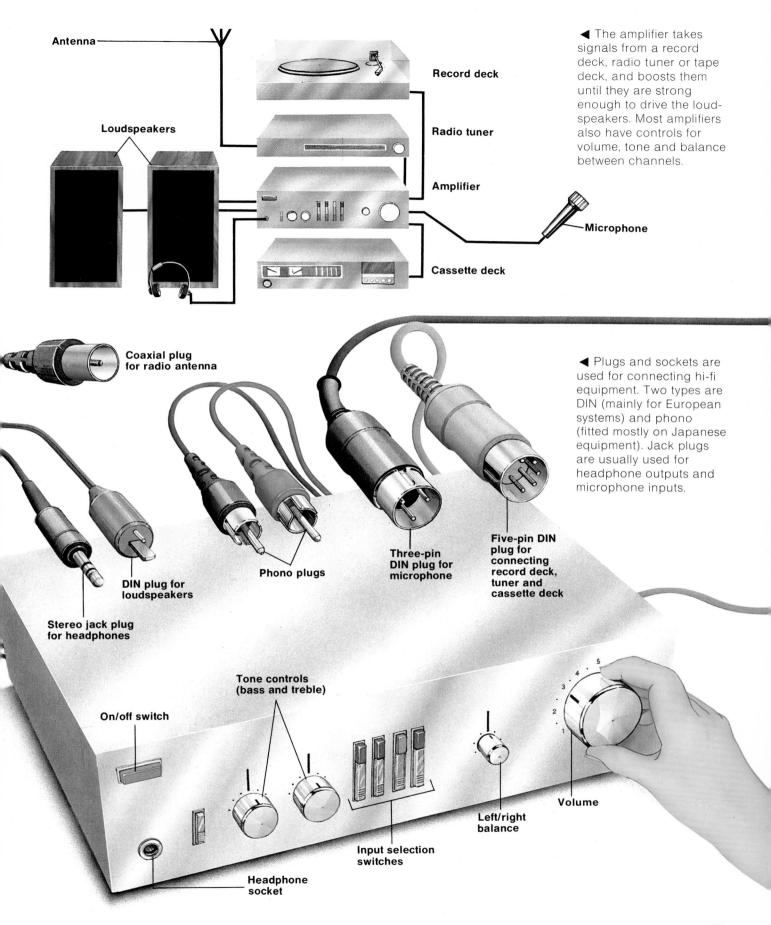

Antenna

Loudspeakers

Record deck

Radio tuner

Amplifier

Microphone

Cassette deck

◄ The amplifier takes signals from a record deck, radio tuner or tape deck, and boosts them until they are strong enough to drive the loudspeakers. Most amplifiers also have controls for volume, tone and balance between channels.

Coaxial plug for radio antenna

◄ Plugs and sockets are used for connecting hi-fi equipment. Two types are DIN (mainly for European systems) and phono (fitted mostly on Japanese equipment). Jack plugs are usually used for headphone outputs and microphone inputs.

DIN plug for loudspeakers

Phono plugs

Three-pin DIN plug for microphone

Five-pin DIN plug for connecting record deck, tuner and cassette deck

Stereo jack plug for headphones

Tone controls (bass and treble)

On/off switch

Input selection switches

Left/right balance

Volume

Headphone socket

Loudspeakers and Headphones

Loudspeakers and headphones are known as transducers: they convert energy from one form into another. In this case, they convert the electrical signals from the amplifier into sound. Speakers must be able to give out a full range of sound from the deepest bass notes to the highest notes we can hear. There are several ways in which they can do this.

Moving-coil or Dynamic Speaker

Moving-coil loudspeakers have a coil of wire attached to a cone. Signals from the amplifier make the coil move, or vibrate. These movements are passed on to the cone which disturbs the air to produce sound waves.

The moving-coil speaker contains special drive units which handle different sections of the sound range. The drive unit reproducing the low-frequency sounds is called a "woofer". The "squawker" deals with the middle range of frequencies, and the "tweeter" copes with high-frequency sounds.

To produce loud noises at low frequencies (for example a bass note below 200 Hz), a large mass of air has to be moved or vibrated. Because air behaves like a fluid at low frequencies, a large radiating surface is needed to create the air disturbance and hence the sound. Therefore, the "woofer" tends to be the largest drive unit in the speaker. High frequencies are not radiated in the same way, and so the transfer of energy from the loudspeaker face to the air is more

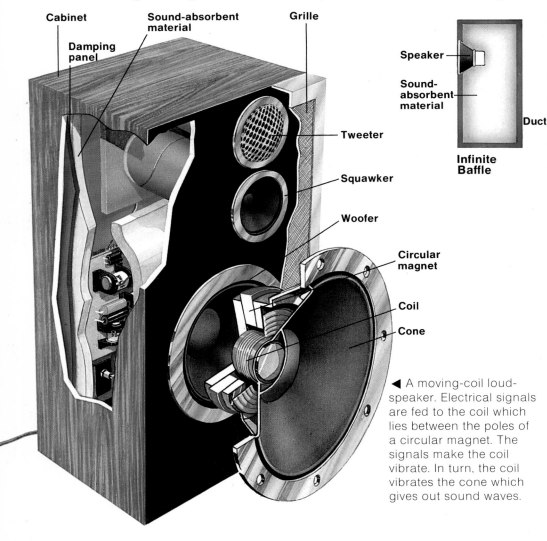

Cabinet

Sound-absorbent material

Grille

Damping panel

Tweeter

Squawker

Woofer

Circular magnet

Coil

Cone

Speaker

Sound-absorbent material

Duct

Infinite Baffle

Bass Reflex

Labyrinth

▲ Three types of loud-speaker enclosure. In the infinite baffle, there is no air path between the rear and front of the cone. The bass reflex allows sound from the rear to reinforce sound from the front. Labyrinth enclosures form a maze.

◄ A moving-coil loud-speaker. Electrical signals are fed to the coil which lies between the poles of a circular magnet. The signals make the coil vibrate. In turn, the coil vibrates the cone which gives out sound waves.

▶ Most headphones work in a similar way to dynamic (moving-coil) loudspeakers. Electrostatic headphones produce sound from a diaphragm which moves to and fro between perforated plates.

efficient. This is why the "squawker" and "tweeter" units are smaller.

Loudspeakers come in all shapes and sizes, often with two, three or more drive units. In recent years there has been a demand for smaller speakers which still offer a good base response and overall sound balance.

Other Types of Speaker

Apart from moving-coil speakers, there are others which use different methods for transmitting sound. For example, there is the piezo-electric speaker. When a signal voltage is applied to it, it moves mechanically to produce sound waves. The ribbon speaker has a corrugated metal ribbon. Signals sent through the ribbon make it move, or vibrate, to create sound waves.

The best-known electrostatic speaker is the Quad system. The first model came on to the market in 1957. The latest Quad speaker is a top

performer, but a pair costs about 40 times more than the average cheap music-center loudspeaker. In electrostatic speakers, sound waves are created by a thin plastic diaphragm, or sheet, which vibrates between two plates. The plates are perforated with holes to let the sound through.

Headphones

Stereo headphones are becoming very popular. They are mostly of the moving coil or electrostatic type. Listening through a headset is a totally different sensation after listening to sound coming from a loudspeaker. This is because the sound enters directly into your ears, giving a strong "presence" or lifelike quality. The best headphones produce high-quality sound and they have the advantage that you can play music as loudly as you like without disturbing anyone else in the room, or the neighbors.

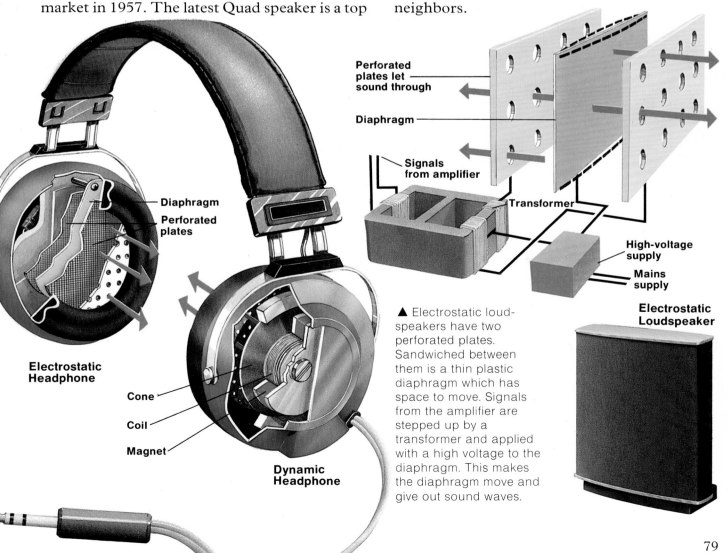

Perforated plates let sound through

Diaphragm

Signals from amplifier

Transformer

High-voltage supply

Mains supply

Electrostatic Headphone

Diaphragm
Perforated plates

Cone

Coil

Magnet

Dynamic Headphone

Electrostatic Loudspeaker

▲ Electrostatic loudspeakers have two perforated plates. Sandwiched between them is a thin plastic diaphragm which has space to move. Signals from the amplifier are stepped up by a transformer and applied with a high voltage to the diaphragm. This makes the diaphragm move and give out sound waves.

Recording at Home

Tape recorders, whether open-reel recorders or cassette players, record and play back sound in the same way. Sounds are turned into electrical signals by the microphone. These signals are then fed to the recording head of the tape machine which transforms them into a magnetic pattern on the tape. When the tape is replayed, the playback head "reads" the magnetic pattern on the tape and turns it back into electrical signals which are fed to the amplifier.

Open-reel Recorders

The introduction of the open-reel recorder in the 1940s revolutionized home recording. For the first time it became easy to store sound on magnetic tape and then to play it back. It also increased the possibilities for "playing around" with sound.

Open-reel tape recorders have retained their popularity because it is easy to edit the tapes. Broadcasting and recording studios use the more advanced versions. However, open-reel recorders for home use are bulky and fairly heavy, and unfortunately rather expensive.

Tape Cassettes

Tape cassettes have extended the tape revolution. There is now a standard cassette which can be played on any deck. In the early days, the sound quality of the tape was poor when compared to an LP disk. But improvements in the tape coatings, and the introduction of Dolby and other noise-reduction systems, have brought about much higher standards. But to achieve good results you do need a first-class cassette machine.

Microphones

Microphones came in various types: dynamic (moving-coil), ribbon, and the cheaper ceramic or crystal elements. Other designs include the expensive electrostatic and the electret.

Microphones transform vibrations in the air (sound waves) into electrical signals. A moving-coil microphone has a thin plate, or diaphragm, connected to a wire coil which is suspended between the poles of a magnet. The diaphragm vibrates when sound waves strike it. In turn, the diaphragm vibrates the coil. As the coil moves,

Open-reel tape machines give high-quality recordings which can easily be edited by splicing. Tape for cassette machines has improved greatly in quality, and there is now a large range of pre-recorded cassettes available.

Mono Cassette Recorder

Open-reel Tape Recorder

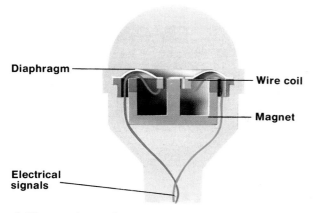

Diaphragm ——
—— Wire coil
—— Magnet

Electrical
signals

▲ The moving-coil (dynamic) microphone works like a speaker in reverse. It contains a wire coil connected to a diaphragm. The coil moves in and out of a magnet when sound waves strike the diaphragm. As the coil moves, it produces electrical signals.

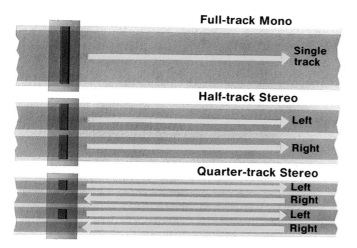

Full-track Mono

Single track

Half-track Stereo

Left

Right

Quarter-track Stereo

Left
Right
Left
Right

it produces electrical signals which vary in strength with the vibrations. The signals are stored on tape as a magnetic pattern.

Using Tape

There is now a vast range of pre-recorded cassettes. But the world is full of interesting people, events and sounds which you can record yourself using a microphone and a cassette machine.

There are several tricks you can use when making your own recordings. For example, sound can be transferred singly from one track to another in synchronization, a process known as "sound on sound". This means that from a simple musical instrument a complete sound program can be built up. On some open-reel machines the tape can be speeded up or slowed down to give interesting sound effects.

Editing Tape

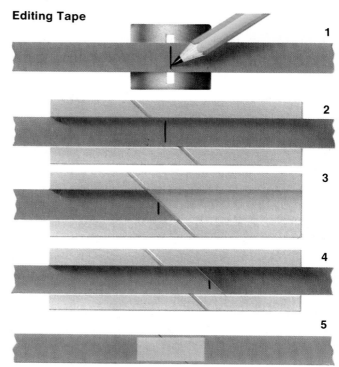

1
2
3
4
5

◄ Domestic open-reel tape recorders use ¼-inch (6.25-mm) tape, and operate at speeds up to 15 inches (38.1 cm) a second. The best type of tape for recording at home is half-track stereo.

▲ Mark the tape at the point to be cut (1). Lay the tape in the groove of the splicing block (2) and cut through it (3). Push the ends of the tape together (4) and put splicing tape over the join (5).

Editing Tape

Tape can be edited to remove unwanted sections, or to put items in the correct order. The best way of editing tape is by splicing. It is possible to re-record from one machine to another to get the items in the correct order, but the skilled use of the razor blade is the most professional way. Unfortunately, this method of editing is not usually possible with cassette machines.

Splicing blocks and proper splicing tape can be bought for this job. Use a chinagraph pencil to mark the section of the tape you do not want. Then cut across the ends of the section with a single-sided razor blade, and remove the unwanted tape. The cuts can be made straight across the width of the tape, or at an angle of 45°. Finally, push the ends together and lay a piece of splicing tape across the join.

Digital Sound

Most of the sound systems described in this book are known as "analog". The electrical signals are continuous and are stored as magnetic variations on tape or as wiggles (waveforms) in a record groove. These signals are a replica, or *analog*, of the sound waves that created them. When replayed, the signals are fed through the amplifier and loudspeakers to be converted back into sound. In such a chain, the sound may be distorted and lose some of its quality.

Digital Audio

In digital recording, the frequency and amplitude of sound waves are measured and converted into electrical pulses. The pulses are coded as a series of numbers, or digits, in binary form. Binary code recognizes only two digits, 0 and 1, which stand for "pulse" and "no pulse".

When a digital recording is played back, the signals are decoded into the usual analog signals which drive the loudspeakers. This system gives a much more accurate result than analog systems because the player "samples" the original recording.

A digital system has to store a great amount of information about the sounds being recorded. It does this in the form of digits, or "bits". As there are far too many bits to be stored on ordinary tape, a video tape or disk is used instead.

The Compact Disk

The Compact Disk is $4\frac{3}{4}$ inches (12 centimeters) in diameter and has a playing time of one hour. The digital information is stored in tiny pits instead of a groove, and the disk is scanned from underneath by a laser beam. There is no contact between the disk and the beam, so there is no wear and tear.

The playing side of the disk is protected by a layer of transparent plastic. With some Compact Disk players you can play tracks in any sequence, and there is a forward and backward search so you can quickly find a particular track.

At present, the digital signals on the disk have to be converted back to analog signals to drive the loudspeakers. A complete digital system may be ready by the end of the 1980s. But the present system, even with its coding and decoding conversions, makes the sounds less prone to distortion, noise and speed changes.

Other Digital Systems

The AHD (audio high-density) disk has a diameter of $10\frac{1}{4}$ inches (26 centimeters), and the signals are stored in a spiral track cut in a PVC material. The tracks are scanned by a special stylus which carries a small electrode. This disk is double-sided and gives an hour's playing time on each side.

A third type of digital disk is the MD, or mini-disk. This is $5\frac{1}{4}$ inches (13.5 centimeters) in diameter, and uses hill-and-dale grooves to store the signals.

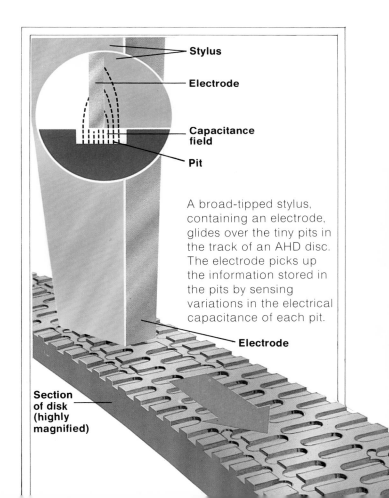

Stylus

Electrode

Capacitance field

Pit

A broad-tipped stylus, containing an electrode, glides over the tiny pits in the track of an AHD disc. The electrode picks up the information stored in the pits by sensing variations in the electrical capacitance of each pit.

Electrode

Section of disk (highly magnified)

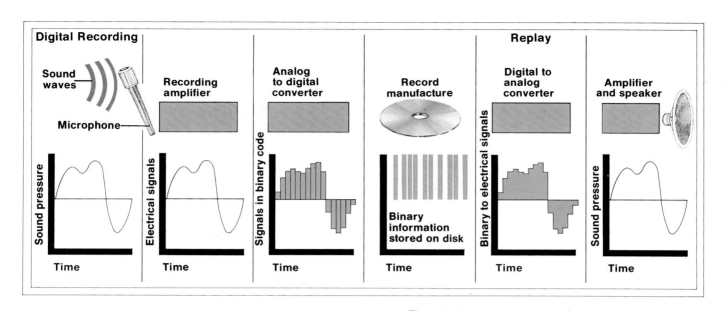

Digital Recording

Sound waves → Microphone

Sound pressure / Time

Recording amplifier

Electrical signals / Time

Analog to digital converter

Signals in binary code / Time

Record manufacture

Binary information stored on disk / Time

Replay

Digital to analog converter

Binary to electrical signals / Time

Amplifier and speaker

Sound pressure / Time

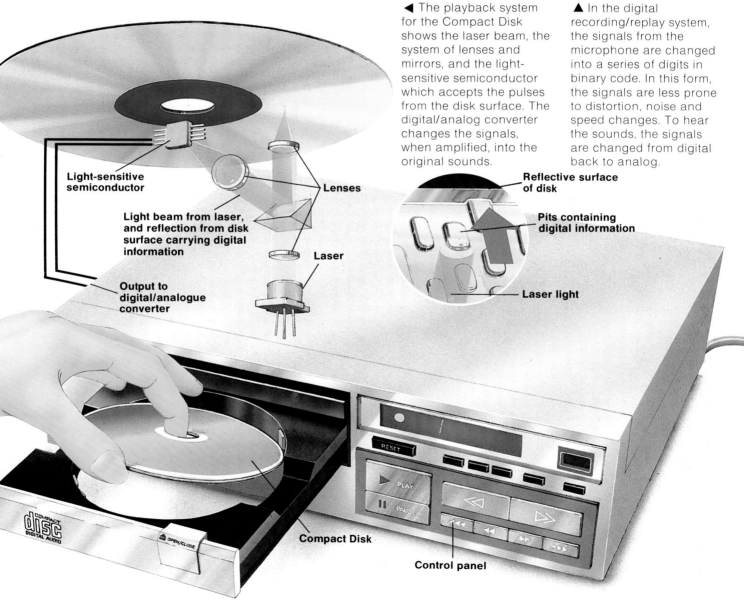

Light-sensitive semiconductor

Light beam from laser, and reflection from disk surface carrying digital information

Lenses

Laser

Output to digital/analogue converter

Reflective surface of disk

Pits containing digital information

Laser light

Compact Disk

Control panel

◄ The playback system for the Compact Disk shows the laser beam, the system of lenses and mirrors, and the light-sensitive semiconductor which accepts the pulses from the disk surface. The digital/analog converter changes the signals, when amplified, into the original sounds.

▲ In the digital recording/replay system, the signals from the microphone are changed into a series of digits in binary code. In this form, the signals are less prone to distortion, noise and speed changes. To hear the sounds, the signals are changed from digital back to analog.

In the Recording Studio

When it comes to making a record, there are two options. The recording can either be made "live" on location (in a concert hall for example) or it can be made in a recording studio. Most recordings are carried out at special studio sessions where there is no audience. If someone makes a mistake it is easy to record that part of the performance again. It is also possible to adjust the balance among the various instruments and voices.

A Classical Recording Session
To record a full-sized orchestra, perhaps with a chorus, a very large studio is needed. Apart from the performers there are also the recording technicians, led by the session producer and the balance engineer.

At one end of the studio there is usually a double-glazed window. Behind this is a soundproof room containing the tape machines, amplifiers and a variety of electronic equipment. The centerpiece of this control room is the mixing desk, which handles the sound levels of a number of microphones or other inputs. Each input is linked to a channel, which in turn will have a fader to adjust the sound level. On the big desks, called consoles, other knobs modify the signals and route them to various tape tracks, monitor positions and monitor loudspeakers.

Recording a Pop Record
When a pop group records a disk, every bit of technical gadgetry is used. This may involve multi-track taping where there is a voice on one track, a piano on another, drums on a third, and so on up to 24 tracks on a 2-inch (50-millimeter) tape. But 1-inch (25-millimeter) tapes are now widely used for the original recording.

The technical tricks used in pop sessions include overdubbing ("sweetening"). This enables a musician not available on the first day of recording to listen to sounds already on tape and then to add his or her contribution. And when the sound balancing of a multi-track tape is critical, adjustments can be made at a "remix" stage. This entails editing down all the tape to a master mix tape which will end up as the LP disk or single.

▲ A close-up of part of a mixing console. This handles 36 inputs, giving 24 outputs which are mixed down to twin-track stereo.

▶ Rock musicians in the studio. They listen to the other instruments through headphones. Acoustic screens separate the performers to avoid sound "spillage".

◀ The producer and engineer sit at the mixing console. They listen to the orchestra (seen through the sound-proof window) through loudspeakers. The sounds are monitored at the console and then fed to the tape-recording machine.

How Disks and Cassettes are Made

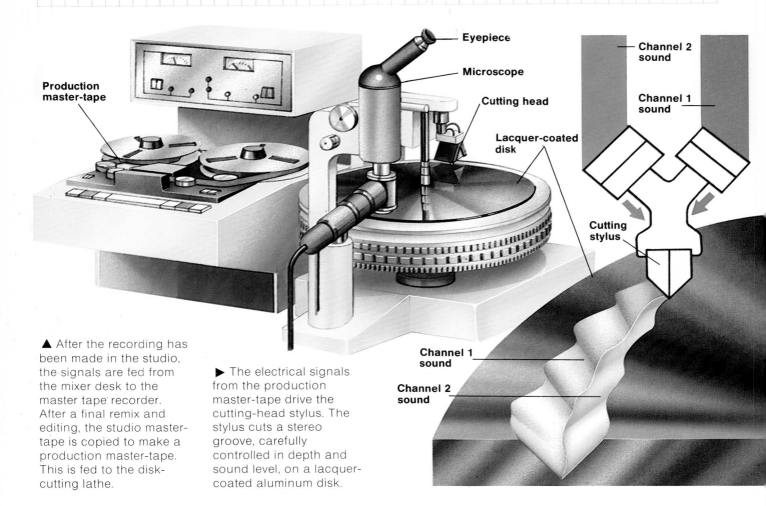

Production master-tape

Eyepiece

Microscope

Cutting head

Lacquer-coated disk

Channel 2 sound

Channel 1 sound

Cutting stylus

Channel 1 sound

Channel 2 sound

▲ After the recording has been made in the studio, the signals are fed from the mixer desk to the master tape recorder. After a final remix and editing, the studio master-tape is copied to make a production master-tape. This is fed to the disk-cutting lathe.

▶ The electrical signals from the production master-tape drive the cutting-head stylus. The stylus cuts a stereo groove, carefully controlled in depth and sound level, on a lacquer-coated aluminum disk.

After a recording has been made at a studio, the master tape is re-recorded on to a lacquer-coated blank disk. This is done on a disk-cutting machine. The blank disk is placed on the turntable of the machine and rotated at a very accurate speed. This is usually $33\frac{1}{3}$ or 45 r.p.m. (revolutions per minute). The cutting head of the machine travels across the disk from the outer edge towards the center, cutting a spiral-shaped groove. The cutting stylus is V-shaped and is made from artificial sapphire or sometimes real diamond.

Compression Molding

The lacquer disk is given a coating of silver and then placed in an electroplating bath for several hours to coat it in nickel. When the nickel-plated disk is stripped from the lacquer it becomes a "master" with ridges equivalent to grooves. The master is coated again to form a positive, or "mother". A "stamper", or metal negative, is electro-formed from the positive and used as a mold to make pressings of the final disks.

After careful centering of the stamper to drill the center spindle hole of the disk, the final pressing is made. One stamper for each side of the record is placed in the jaws of a heated press. The labels are placed in the press with a lump of vinyl plastic, which has been preheated. The pressing time is usually less than 30 seconds. Cooling water is then passed through the press, the jaws are opened, and the disk can be extracted by the operator.

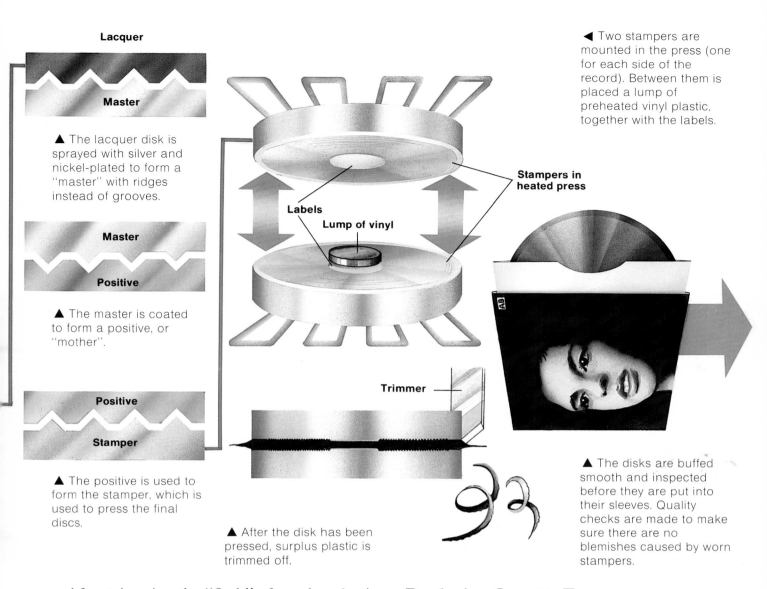

Lacquer

Master

▲ The lacquer disk is sprayed with silver and nickel-plated to form a "master" with ridges instead of grooves.

Master

Positive

▲ The master is coated to form a positive, or "mother".

Positive

Stamper

▲ The positive is used to form the stamper, which is used to press the final discs.

◄ Two stampers are mounted in the press (one for each side of the record). Between them is placed a lump of preheated vinyl plastic, together with the labels.

Stampers in heated press

Labels

Lump of vinyl

Trimmer

▲ After the disk has been pressed, surplus plastic is trimmed off.

▲ The disks are buffed smooth and inspected before they are put into their sleeves. Quality checks are made to make sure there are no blemishes caused by worn stampers.

After trimming the "flash" of surplus plastic left round the disk, it is buffed smooth ready for placing in its sleeve. This method is known as compression molding.

Other Methods

Other methods of making disks include injection molding and embossing. There is also the direct-cut disk where the lacquer blanks are cut without using an intermediate tape stage. The aim is to remove any distortion or noise, but this system does mean that the performers cannot make any mistakes, for no editing is possible. Compact Disks are pressed in a similar way to vinyl disks, but the process used for producing the master disk is more refined as each disk has billions of tiny pits on its surface.

Producing Cassette Tapes

Pre-recorded cassettes are now so popular that major record companies often issue disks and cassettes simultaneously. The mass production of tape is a different and more expensive system than disk processing.

The master tape, usually running at $7\frac{1}{2}$ inches (19 centimeters) a second, is transferred to a special 1-inch (25-millimeter) wide master tape. This has four tracks for two stereo channels joined into a loop. The tape is loaded into a special unit running at high speed. Its sound signals are fed to a dozen or more "slave" machines which copy the recording. The recorded reels are then placed in a machine which automatically cuts the copies to the right length, and feeds the tape into cassette boxes.

Radio Broadcasting

During the early years of television, it was feared that the days of radio were numbered. Since then, radio has undergone many changes. But today, because of its attention to special interests such as music and education, many people still listen to it. Local radio stations, which serve their own districts, are also growing in number. They are particularly useful in supplying local news.

How Radio Works

When sound, or audio, signals are broadcast, they are combined with a high-frequency wave known as the carrier wave. The high frequencies of the carrier wave are referred to as radio frequencies (RF). The terms used to describe these are Megahertz (MHz) and Gigahertz (GHz). One MHz equals 1,000,000 Hz, and one GHz equals 1,000,000,000 Hz.

The process of putting the sound signals on the carrier wave is called "modulation". This can be done in several ways. The most widely used in sound broadcasting are AM (amplitude modulation) and FM (frequency modulation). The first method has been used since the start of broadcasting.

Broadcasts on short wave (SW), medium wave (MW) and long wave (LW) are AM transmissions. In these, the amplitude of the radio signal varies according to the strength of modulating sound from the microphone, disk or tape. After dark, MW transmissions are often not well received because of interference from distant stations. With FM broadcasts, the frequency of the carrier is made to vary up and down according to the audio signal.

Stereo Broadcasts

For stereo radio transmission, two separate audio signals (left and right channels) are sent out by one transmitter. The two channels are coded to form a "multiplex" signal. This is frequency-modulated on to a standard carrier. To grasp roughly how the system works, think

▶ How radio waves travel. Short waves shoot upwards and are reflected down by the ionosphere. VHF waves can be sent over the horizon by high transmitting and receiving antennas. Super-high frequency radio waves shoot out into space but can be reflected back by satellites.

◀ A radio interview with Elton John. The mikes have windshields to avoid popping and breath noises when talking close up. The interviewer receives cues through his headphones from the producer in the control room.

▼ Reading the news. The newsreader receives instructions from the control room through his headphones.

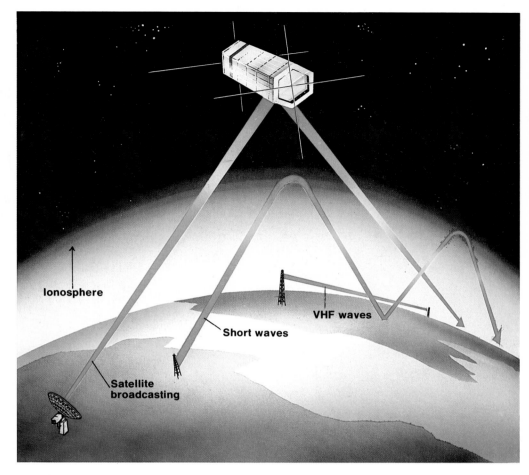

Ionosphere

Short waves

VHF waves

Satellite broadcasting

of the left and right signals switching alternately at very high speed (38,000 times a second) on to the carrier wave. A "pilot tone" keeps the transmitter and radio receiver synchronized. The system is "compatible", which means that an ordinary radio set can pick up mono sound from a stereo transmission.

Hi-fi enthusiasts always use VHF (very high frequency) wherever possible. The FM technique needs an effective antenna and care in tuning the dial to the required station. On a press-button radio, the pre-set tuning must be done accurately.

Outside Broadcasts

To make programs from locations outside the studio, an outside broadcast (OB) van or a radio car is needed. These have facilities for receiving and mixing sounds from various sources. The signals are transmitted back to base on UHF (ultra-high frequencies), with cues and instructions passed over a two-way VHF channel.

Music on the Move

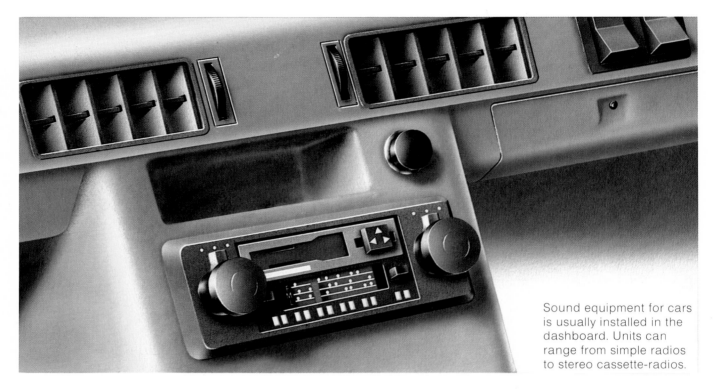

Sound equipment for cars is usually installed in the dashboard. Units can range from simple radios to stereo cassette-radios.

Portable hi-fi equipment is becoming increasingly popular, particularly with the introduction of pocket-sized cassette players and stereo cassette-radios. Many of these can be run on batteries or plugged into the mains.

Mobile sound systems have also become a standard fitting in new cars. The inside of a car, however, is not the best place to listen to hi-fi sound. The acoustics are not good and the temperature variations cause problems for the equipment. These difficulties have been tackled in various ways by different manufacturers.

The Car Radio

Despite recent improvements in the quality of tapes, many sound enthusiasts believe that radio will continue to be the most used signal source in car audio, although Compact Disk players are promised for the future.

Designers of radio equipment for cars have to bear in mind that a motor vehicle is moving, and so radio circuits have to deal with every reception problem encountered when driving. The radio must be easy to operate while driving, so today's systems have push-button station selection as well as manual tuning. Scanning across the tuning band is done automatically, stopping when a strong signal is received.

There are also circuits to reduce noise and to maintain the stereo signal as long as possible. When stereo reception is too poor, the circuit changes to mono sound. Some designs reduce increased noise when going through a tunnel, and improve fringe-area reception. Noise in the VHF/FM band, caused by the ignition system of your car, or "pulse noises" from other vehicles, has to be suppressed by devices in a special circuit.

The latest developments in car radios include digital synthesized tuning, with 15 pre-set stations and an auto-seek system. But the ultimate in tuning convenience (other than a spoken command to find a station) is a foot switch which allows the tuner to seek the desired station automatically after muting the sound during the search.

The Antenna

The one component that will not be found inside any car is the antenna. If it was set inside the car, the signal would fade when the car changed direction. The body metalwork would have a screening effect, and if it were placed behind the dashboard the antenna would pick up ignition interference.

Various types of antenna are available: semi- and fully-retractable wing mounted, side-pillar mounted and roof-top mounted. Motorized antennas can be raised or lowered from inside the car. The best length for an FM aerial is between $2\frac{1}{2}$ and $2\frac{3}{4}$ feet (0.75 and 0.85 meters), but for AM reception there is no preferred length; in fact, the longer the better.

The Loudspeakers

The weakest link in the in-car chain is definitely the loudspeaker system. Loudspeakers come in various formats, including sealed two-way parcel-shelf models, with "woofers" within the rear doors of the car.

The Cassette Player

Until recently, there were two rival tape systems for cars, which required different machines to play them. These were the cartridge endless-loop device and the familiar tape cassette. However, cassettes have taken over because they require far less space inside the car for storage. Cassettes can also be played at home on a cassette deck. With the coming of micro-cassette systems, even smaller all-in-one radio and tape players will soon be available.

To get the best sound reproduction, car cassette players may have features which include noise-reduction systems such as Dolby, separate bass and treble tone controls and a tape selector switch for normal, chrome or metal tapes.

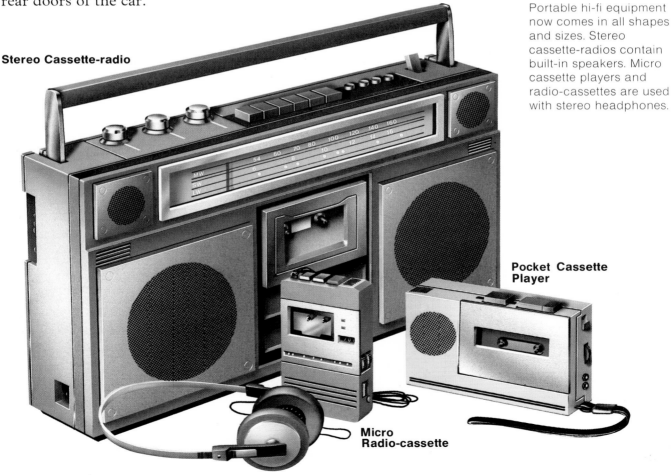

Stereo Cassette-radio

Pocket Cassette Player

Micro Radio-cassette

Portable hi-fi equipment now comes in all shapes and sizes. Stereo cassette-radios contain built-in speakers. Micro cassette players and radio-cassettes are used with stereo headphones.

Radio Communications

In the late 1940s and early 1950s receiving and transmitting equipment became more readily available. Today, anyone can buy a transceiver, and all you need to do to operate it is to erect a simple dipole antenna or buy a directional array and set it up on a roof.

If you want to have a recognized call sign, you have to pass a simple examination.

Citizens' Band Radio

Citizens' Band radio was being developed in the USA as long ago as 1948. It has now spread to more than 60 countries. The basic purpose of CB was to allow amateur radio enthusiasts, who became known as "breakers", to install in their car, caravan, boat or home a modest two-way rig

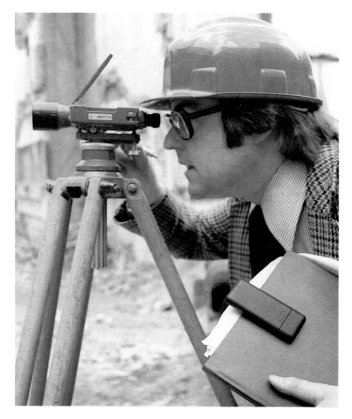

▲ This surveyor carries his radio-pager clipped to a file.

► This truck-driver can communicate with other moving vehicles via his CB set.

◄ Car phones are useful for keeping in touch with the home base.

on which they could discuss road conditions, the weather or other matters with fellow CB users. Unfortunately, "breakers" can cause havoc and interference on certain channels if they do not operate their transceiver correctly. All you need to operate CB radio is a licence that you can buy for a small fee. You do not need any technical knowledge, nor do you need to pass an examination. In many parts of the world CB radio uses AM systems.

Mobile radio has to compete for air space with television, broadcast radio, maritime and aeronautical services, satellite communications and many other uses.

Portable Equipment

Mobile radio services, for private use only, usually have two-way VHF or UHF radio-telephone systems (operating between a home depot and delivery trucks, for instance).

Low-power portable radio-telephones can be hand-held or worn on the shoulder. Paging systems may be designed simply for one-way calling, although a talk-back facility is sometimes needed, as in the case of the police force. Many radio-pagers are small enough to be slipped into an inside pocket.

You can also get personal car phones which allow you to dial direct into the telephone network, so you can call someone when you are driving along.

Another wide use of mobile radio is in radio-controlled models. This use of radio has become highly sophisticated, and the modeler can now accurately simulate the movements of full-sized machines such as aircraft. Radio control relies on the conversion of movements of control sticks, switches or levers into radio signals. These radio signals are re-converted into mechanical signals in the model. The signals are transmitted on such low power that models only 2 miles apart do not interfere with each other.

ROBOTS

The word "robot" has its roots in fiction, but today there are thousands of robots at work throughout the world. This section looks at the impact robots are making on our lives. It shows how they work, how they can be equipped with "sense" and "intelligence", and what the future holds in store for them.

How Robots Work

Industrial Robots

Robots in Fiction

Robots of the Future

Introducing Robots

Android

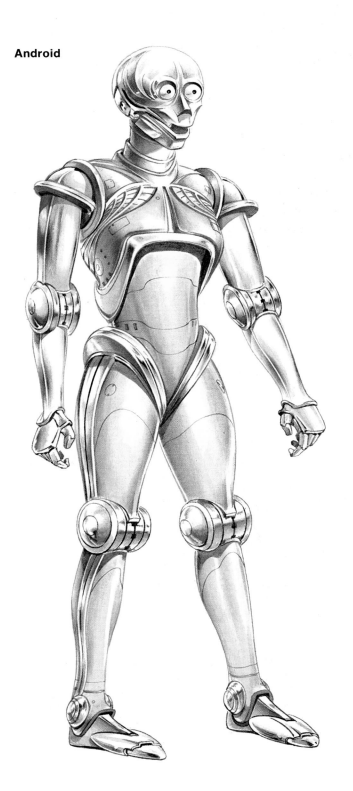

For many years robots have excited people's imaginations – whether in books, in films or inside fairgrounds or amusement arcades. But it is only quite recently that robots have left the realms of fiction. They are becoming useful to men and women in all sorts of ways. How this change has come about, and how it will progress over the next few years, is the subject of this book.

Even now thousands of robots are at work around the world. Their activities range from delicate construction work to heaving huge loads, and they can be found anywhere from factories and nuclear power stations to outer space. In the future robots may become an even greater part of our everyday lives. With the help of "brains" that are in fact powerful computers, robots may become as clever as people. They could take over a wide variety of jobs – in offices, in factories or in the home. Robots could also work in particularly unpleasant places where complex, vital work needs to be done, such as mines or the bottom of the ocean.

What is a Robot?

But first, what exactly do we mean by "robot"? The word can be defined differently depending on who you talk to, or in which part of the world you happen to live. The widest possible definition of a robot is "a machine which imitates a person in appearance or action". Note the emphasis on some kind of physical likeness to people. Over the years, robot-makers have been especially interested in designing robots which, in some way, look like men or women. It is this

◀ For many years people thought of robots as devices that looked and behaved like humans – androids or mechanical men. Nowadays robots have been redefined to mean mechanical àrms that are controlled by a computer.

▶ One of the best-selling industrial robots is the Unimate, made by the American firm Unimation. This machine can do a variety of jobs depending on how it is controlled. It can weld metal or move objects from place to place.

which has given robots their superhuman reputation. People regard them with a mixture of fascination and fear. The idea that a machine can look like a person, or faithfully copy what he or she does, summons up feelings ranging from wonder and amazement to downright distaste.

Androids and Automatic Devices

The robots we are most familiar with are the ones from science fiction that faithfully take on the appearance and actions of men or women. Such devices, which one might call the most advanced forms of robot, should properly be called "androids". The word comes from the Greek words *andros* meaning "man" and *eidos* meaning "form".

Dolls that walk and talk, or do other things to imitate people, can also be called robots. Many people, however, confuse true robots with what are really simple automatic devices. Gadgets such as clocks or traffic lights work automatically, but they are not true robots.

Robots Go to Work

For many years engineers tried to find jobs for robots to do. But, outside amusement halls or the world of films, this proved very difficult. The reason was straightforward. Engineers lacked the skills, and the essential tools, to make robots reliable enough to do jobs in real life.

Two developments changed all this. First, engineers in the United States set their sights lower and decided that robots should do no more than imitate people. They need not necessarily look like people at all. To engineers, robots have simply become mechanical contraptions that imitate the actions of human arms, and are controlled by computers. The second development is that engineering skills, particularly in computers, have progressed at a great pace. The world has become better at making machines that fit the new definition of what a robot is. The result is that, almost overnight, robots have left the kindergarten stage. They have started to go to work.

Industrial Robot

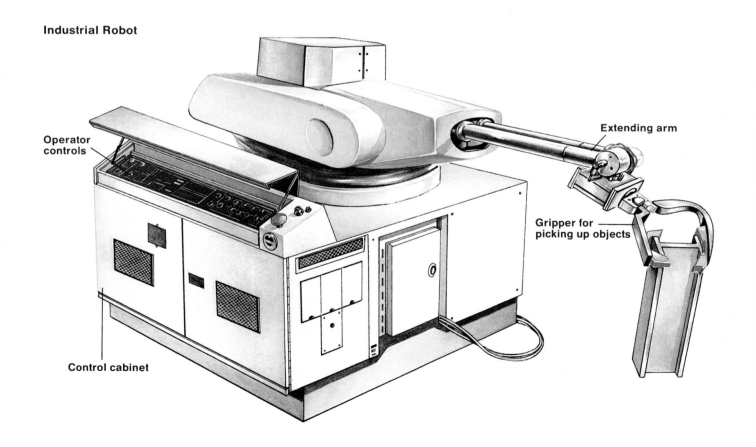

Operator controls

Extending arm

Gripper for picking up objects

Control cabinet

Anatomy of a Robot

The human body contains two sets of components, or parts, connected with action. The first is made up of the brain and nervous system, and the second of limbs and the other parts of the body that move. Robots are made up of components with similar functions. But the average robot at work today is nothing like as complicated as a human being. For one thing, its range of actions is limited by it having only one limb – a single "arm". The instructions that control it, however, are relayed in a similar way to the messages that the brain passes on to the rest of the human body via the nervous system.

In a robot, the job of a "brain" is done by a computer. The computer may store information obtained from the outside world by sensors on the robot. These sensors take on a role similar to eyes and ears in humans. But most robots today are not advanced enough to have sensors. They resemble people who are not only blind and deaf but also lack any way of obtaining information directly from the outside world.

Programming and Electrical Messages

Most robots, then, must receive information in a way that has no direct parallel in humans or other animals. They receive instructions by the mechanism of programming. To program a robot, a human operator feeds the computer messages about the kinds of jobs he or she wants the robot to do.

The really important point about robots is that, by giving a robot's computer a series of different programs, an operator can make the robot do a wide range of jobs. For example, the operator could feed in details about cleaning a window and watch while the robot does it. The next minute, by pressing another button, he or she could tell the robot to pick up a lump of scrap iron. It is this ability to switch quickly from one kind of job to another that makes robots useful in industry.

Like a human being, a robot needs some way to channel the information from its "brain" to

its "actuator" – in this case, its one and only arm. In a human, this channel is made up of a series of nerves. Thousands of electrical impulses stream along these nerves and tell, for instance, a hand to pick up an apple or a foot to kick a ball. In a robot the information channel is made of electrical cables. These drive motors which move the different components in the robot's arm.

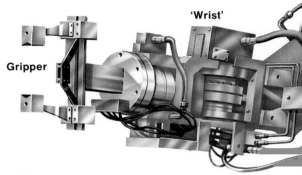

Gripper · 'Wrist'

The Robot Arm

Though it has to be content with just one limb, a modern robot can do quite a lot with its rather limited bodily equipment. A robot arm can position objects to an accuracy of a hundredth of an inch. Some robot machines can lift weights as heavy as one ton. A typical robot arm contains a "forearm", an "elbow" and an "upper arm" as well as a "hand". In other words, it is very similar to a human's. The linkages at the shoulder, elbow and wrist enable the robot arm to move around a total of six axes.

The devices that power the different parts of the arm fall into three types. They rely on the movement of compressed air (pneumatic power), on liquids that move under pressure (hydraulic power) or on the power provided by a series of electric motors.

One of the most important parts of the robot's body is its "hand". This has final responsibility for the jobs that the robot does. Robot hands come in different styles for different tasks. The hand can be a claw, a suction pad, or a magnet. It may also be a special unit into which a tool for a particular job may be fitted.

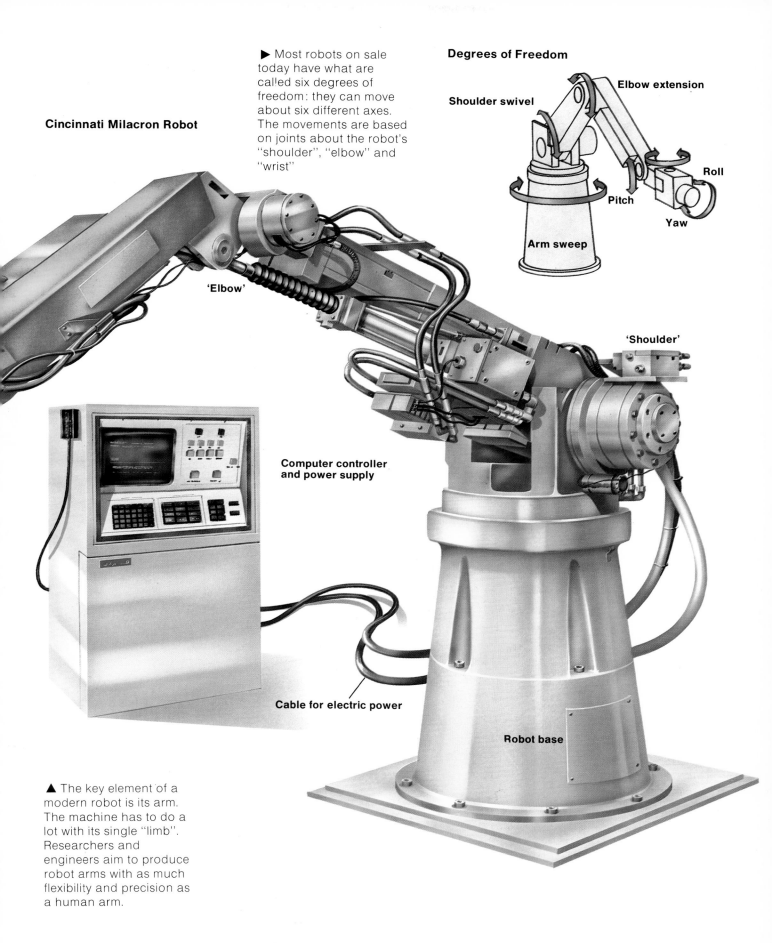

Cincinnati Milacron Robot

▶ Most robots on sale today have what are called six degrees of freedom: they can move about six different axes. The movements are based on joints about the robot's "shoulder", "elbow" and "wrist"

Degrees of Freedom

Shoulder swivel

Elbow extension

Roll

Pitch

Yaw

Arm sweep

'Elbow'

'Shoulder'

Computer controller and power supply

Cable for electric power

Robot base

▲ The key element of a modern robot is its arm. The machine has to do a lot with its single "limb". Researchers and engineers aim to produce robot arms with as much flexibility and precision as a human arm.

Industrial Robots

There are 40,000 or so robots existing throughout the world today. Of these probably 90 per cent work in factories. There is nothing remarkable about this fact. It is just that robots are very good at doing straightforward jobs which involve lifting or some other kind of arm action. The most obvious places where robots can take over these kinds of jobs is in factories.

The Limitations of Factory Machinery

Factories are places in which people make large quantities of goods. Until the Industrial Revolution, which was at its height during the late 18th and early 19th centuries, people made things, but not in a highly organized fashion. Nor did they use machines to do much work for them. Since the Industrial Revolution both the level of organization and the use of machinery have been growing steadily.

So there is nothing new about machines doing the work of people. Until recently, however, most machines in factories did repetitive jobs, or jobs in which skill with the hands and fingers was not involved. For example, an automatic machine might be very good at putting pins through a number of holes in a metal plate. But the pins must always be the same size and shape, and the holes must always be in the same position in the plate. The machine would not be able to vary the job in any way; for instance, by putting a different shape of pin into the holes.

To take another example, machinery in factories has been used for years for channelling liquids through valves and for monitoring their temperature. Such processes are important in many types of industry where the chemical composition of liquids or solids is changed. These industries range from oil refining and steel-making to the processing of foodstuffs.

▼ Here a robot is picking up a casing for a refrigerator and feeding it to a trimming machine.

▼ A common job for factory robots is spraying paint. The machine's arm can move in a way similar to a person's to follow a complicated pattern and paint irregularly-shaped objects. The robot is carefully programmed to do this.

▶ Many industrial products need holes drilled in them. People can do this with tools; but the work is often quicker and more precise if left to a robot. The parts to be drilled are brought to the robots along a moving conveyor belt.

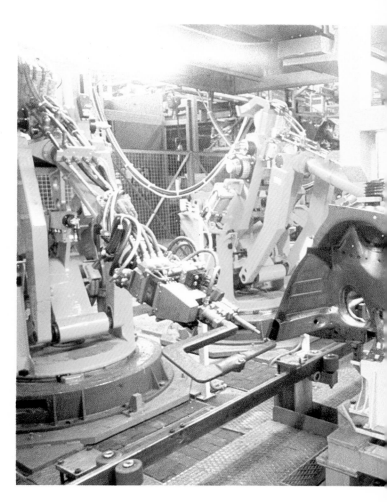

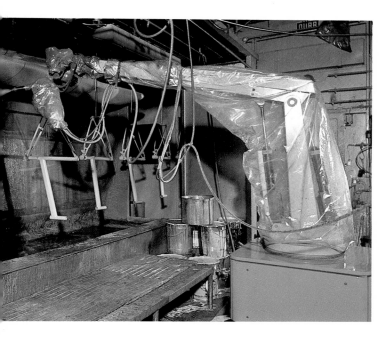

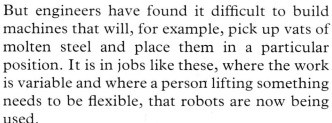

But engineers have found it difficult to build machines that will, for example, pick up vats of molten steel and place them in a particular position. It is in jobs like these, where the work is variable and where a person lifting something needs to be flexible, that robots are now being used.

The Advantages of Robots

Two of the most popular jobs for robots are welding and painting. Robots on car welding lines are a common sight in most parts of the industrial world. Robots are also good painters. Armed with a painting gun, they can be programmed to coat parts that have complex shapes. They can do this more accurately than a human being can. Spray-painting is often a very unhealthy occupation for people, as the paint can get everywhere and irritate the eyes, ears and nose. For such jobs, robots have advantages over people in terms of health and efficiency.

Another job done by robots is the simple lifting of objects from one place to another in a factory. These can be anything from machine parts to bags of cement. A robot will, for instance, stack the objects neatly according to a set program of instructions. If the factory manager wants them stacked differently, or another set of objects stacked somewhere else, all he or she has to do is change the program.

Robots Working with Other Machines

Robots are also used to fit parts such as electronic components into other pieces of equipment, for example printed circuit boards for electrical equipment. They are good at taking objects out of automatic tools, such as machine tools that cut things, or die-casting machines used in molding. In these jobs, the robot is a servant to another type of machinery and is doing a job which would normally be done by a person.

Robots Diversify

Robots are already starting to move out of factories and into a variety of other places. One job which has always been dangerous for human workers is the inspection and maintenance of nuclear power stations. Robots are now beginning to take over this work.

The central part of a nuclear power station is the reactor core. It is here that a nuclear reaction takes place, generating enormous heat. In a conventional nuclear power station, the heat is converted into electricity. Inside the core are rods of radioactive uranium that provide the fuel for the nuclear sequence.

As the fuel is used up, the uranium rods have to be replaced. Ordinary machines have no trouble pulling the rods in and out of the core, but other jobs have to be done. For instance, the rods must be inspected for faults once they are inside the core. Emergency welds on the interior of the core's wall must sometimes be done.

For some years, mechanisms other than robots have done the essential jobs inside reactors. These are called telechiric devices. Telechiric devices are often confused with robots. They are mechanical arms that respond to the instructions of a human operator. In a nuclear power station, the operator would be sitting out of harm's way in a control room on top of the core. The essential feature of a telechiric mechanism is that a human, rather than a computer, is in control of the arm.

Now, however, a new generation of robots is appearing which is suited for work in nuclear power stations. Taylor Hitec, a company near Manchester, England, is working on a mechanism that can slide through one of the small entrance holes in the core. It then unfolds rather like an umbrella, to reveal a robot hand. A computer above the core, to which the hand is linked by cable, programs the machine to do a particular job. This may be a simple inspection job (when the arm holds a television camera) or removing pieces of metal.

Taylor Hitec is also working on another, more ambitious, project. This is a robot that can take apart the highly radioactive core of a power station when it is no longer required. This is too difficult and dangerous a job for people.

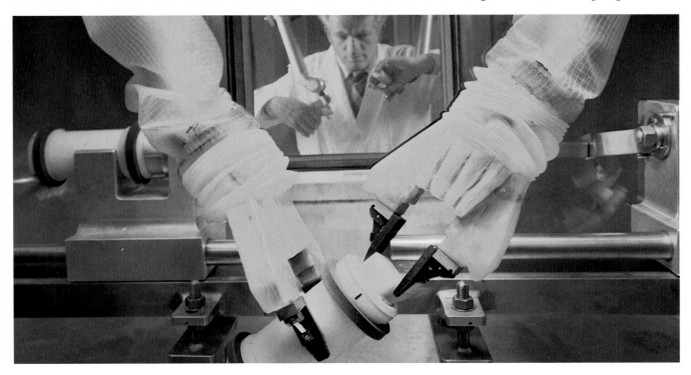

Candy, Underpants and the US Navy

Robots are doing jobs in other unusual areas. Candy firms, for example, use them for putting confectionery into boxes. This is a job normally done by humans but it is boring and can be done more cheaply by a programmable machine.

One large clothing company which makes underpants, wants to put robots to work in its sewing-machine rooms. The robots would grab pieces of material and pass them to sewing machines which would stitch them into the completed garments.

A company called United States Boosters has a design for robots which can shoot high-pressure water to clean up the rocket motors of the Space Shuttle. This would allow the motors to be reused. Also in the United States, a US Navy robot takes out damaged rivets from the wings of aircraft. The US Navy has also designed a programmable mechanism that crawls over the surface of large vessels, cleaning barnacles off them.

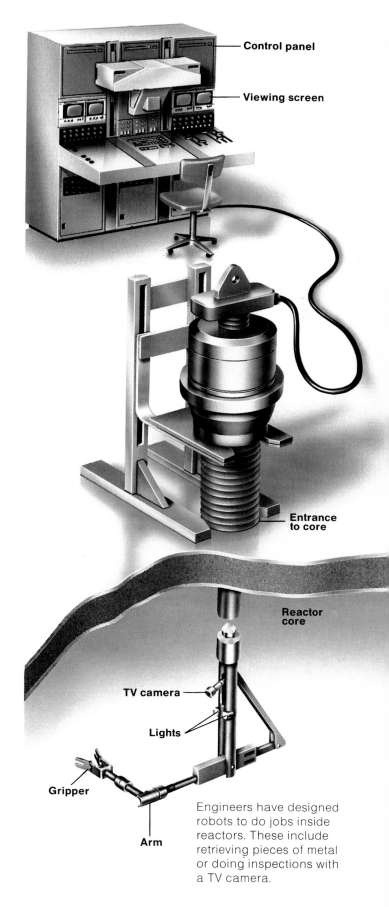

Control panel

Viewing screen

Entrance to core

Reactor core

TV camera

Lights

Gripper

Arm

Engineers have designed robots to do jobs inside reactors. These include retrieving pieces of metal or doing inspections with a TV camera.

◄ Here a technician is using a telechiric arm to do a complex scientific job. The mechanism merely acts as an extension of the person's arm.

▲ Putting candies in boxes is an irksome, task for people. They can put candies in the wrong position or squeeze them too hard, damaging them. Robots are being called in to take over the job.

The Automated Factory

◄ In this automobile factory, car bodies move along a conventional production line and are welded as they pass. A robot welding-line, such as this, is controlled by a central computer. It may produce as many as 1000 car bodies a day.

► In the factory of the future, the three main areas of work (planning and administration, design and manufacturing) will be controlled by the same series of computers. There will be far fewer human workers than there are in today's factories. Most of the workers will be needed for changing the programs in the computers.

To understand the impact of robots on our lives, we should look a little more closely at how they link with other, new types of machinery. Robots by themselves, no matter how "clever" they become, will never do all the jobs inside a factory. They are just one part of a mass of computer-controlled machinery that is gradually becoming more important in industry. This is reducing the need for human workers.

The Beginning of Automation

The arrival of the electronic computer in the 1950s brought a new word into our language – automation. This is the name given to the replacement of human effort by machinery that is controlled by computers. Robots, therefore, are one part of the process of automation, but other types of machinery play an important part.

Using Computers and Robots

There are three main areas of work in a typical factory: planning and administration, design and manufacturing. In planning and administration, the office staff keep a check on the number of goods ordered by customers. They also check the quantity of different parts delivered to the factory by suppliers. Computers play a large part in this work. By using computer terminals, such as display screens and keyboards, workers can gain access to a large computer that contains information about the factory's performance. They can also feed in information that will be useful.

It is a similar story in the design department. Draftsmen can work out the shape of new products on computer terminals. These include screens on which drawings and other informa-

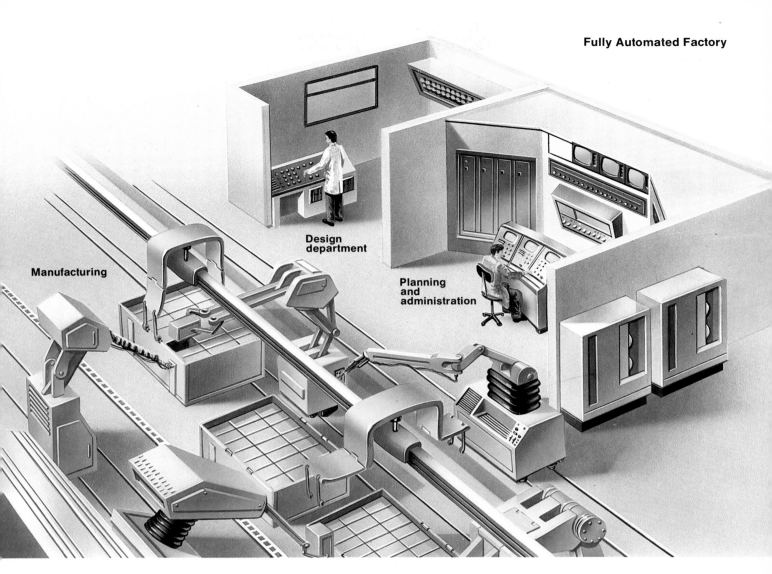

Manufacturing

Design department

Planning and administration

tion can be displayed. The draftsmen can also use their terminals to gain access to essential information stored in the central computer. This might include details of existing products or the factory's expected orders for the next six months.

Robots, as we have already seen, can do many useful jobs in the manufacturing process. They can also work with the machine tools that do cutting and shaping operations on the metal, wood or plastic from which the factory's products are made. These machines are also usually controlled by computers. Programs instruct the machines to cut a piece of metal in a certain way, to drill so many holes at particular points in the material, and so on. In some advanced factories, one computer controls a batch of machine tools and the robots that feed parts into them. The computer may also control other robots that, for example, lift parts from a moving conveyor belt.

Factories of the Future

If these three main areas of factory work can be brought together under the control of the same series of computers, then the factory is about as fully automated as possible. The main feature of such a plant is that few workers are involved in manual operations. Most are in planning and design, plus a few maintenance engineers to keep the machines running smoothly. Most of the workers would be involved in changing the shape and nature of the factory's output by changing the programs in the central computers. Very few factories of this kind exist today, but over the next twenty years they will steadily grow in number.

Teaching Robots

All robots are controlled by computers. But unless a robot has sensors so it can receive information directly from the outside world, someone must put instructions into the computer so that it knows what to tell the robot arm. This is known as programming the robot.

Programming

The most important parts of a computer are the central processing unit and the memory. Instructions in the memory are fed into the central processing unit where they act upon other information (which originally was also stored in the memory). In an ordinary computer, an operator feeds in a set of figures, followed by a set of instructions telling the machine to do a particular operation on the figures. This may be adding up the figures, for example. This set of instructions is called a program.

Computer Codes

The instructions, plus all other information inside the computer's memory, are coded as a series of electrical impulses, each of which can be either "on" or "off". This on/off pattern of pulses can in turn be represented by a sequence of digits in which 0 stands for "off" and 1 stands for "on". Such a pattern of numbers, in which only two digits are used, is called a binary system.

Humans, however, would find it far too complicated to feed information into a computer using the binary system. Computer engineers have therefore worked out special codes that automatically translate instructions using English words into the binary system that computers understand. It is not possible yet to use ordinary sentences. Instead, words are fed into computers in the form of special computer 'languages'.

Programming a Robot

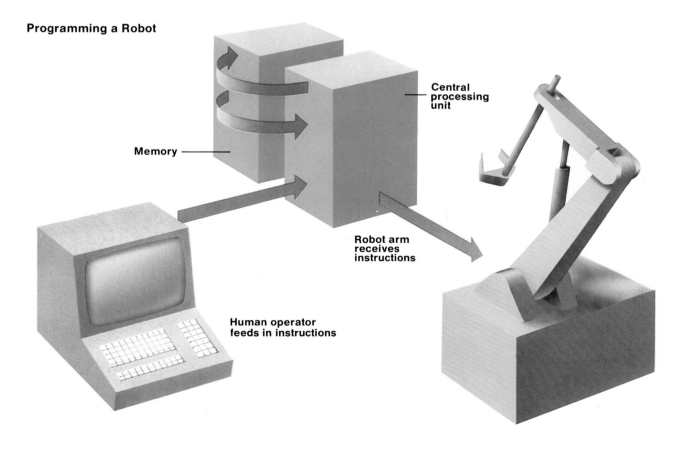

Central processing unit

Memory

Robot arm receives instructions

Human operator feeds in instructions

Data

To program a robot, a person normally needs to give it two sets of information, or data. The first relates to the situation in which the robot must work. For instance, the programmer needs to tell the robot how far away it is from a machine tool (if the job is feeding parts into the tool) and details about the shape and characteristics of the tool.

The second set of data concerns the exact job that the robot is required to do. For the above example, this data would include instructions about the number of parts to feed into the machine and the shape of the parts. This set of instructions could also include extra information that the robot could use if something went wrong. For example, if the supply of parts being fed to the robot on a conveyor belt stops for any reason, then the robot could be told to halt operations.

Also included in the program may be the positions that the robot arm is required to take up as it moves through the handling routine. Such information may be necessary to stop the arm bumping into objects while it is doing its job.

Teaching by Doing

A very useful way of instructing robots is called "teaching by doing". In this method a worker guides the robot through a particular sequence of actions. The robot "remembers" the positions by automatically putting the information in its own memory, along with details of what to do at each stage. "Teaching by doing" is particularly useful when the job is normally done by a highly skilled person, and where feeding exact details of the work into the robot would be difficult.

▶ Teaching by doing is the name for instructing robots to do a particular operation by guiding them through it manually. Here (top) a person is moving a robot arm in the same way as he would move his own arm when painting an object with a spray gun. The robot "remembers" the positions for later use. Other instructions (for instance, to tell the machine to change the color of the paint at a particular point) can be given to the robot using a keyboard. Finally (bottom) the robot repeats the movements to do the job as instructed.

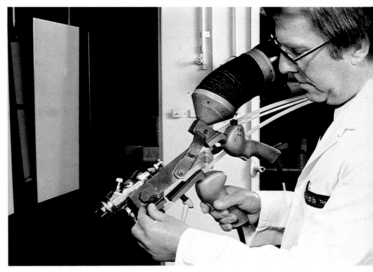

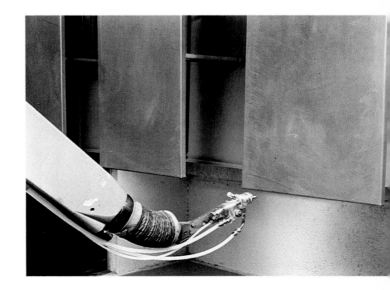

Sensitive Robots

All the robots we have looked at so far have been totally dependent on people. Though the robot may do a job without supervision, someone has to tell the robot's computer what to do by feeding in a program. Also, if something happens during the job that the robot has not been programmed to expect, it is powerless to react.

There is, however, a way around this problem. This is to provide the robot with the kind of mechanisms that people have for receiving information from the outside world and channeling it into the brain. Engineers call these mechanisms "sensors". The most important human sensors are the eyes, ears, nose and nerve endings.

Robots that "See" and "Feel"

The most usual way of giving robots sense is through vision. A television camera takes pictures (perhaps of an object that the robot is supposed to pick up) and feeds the information to a computer. This computer is normally separate from the one that controls the robot's

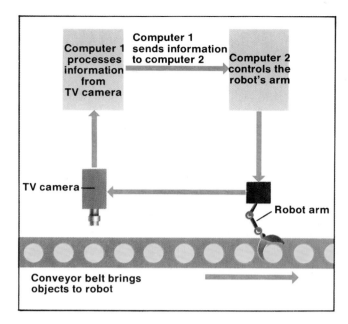

▼ Few robots with vision operate in the world's factories. Here one of them goes through its paces, "seeing" objects with a TV camera so it can pick them up.

▲ A "seeing" robot needs two computers. One computer controls its arm and the other computer makes sense of information provided by a TV camera.

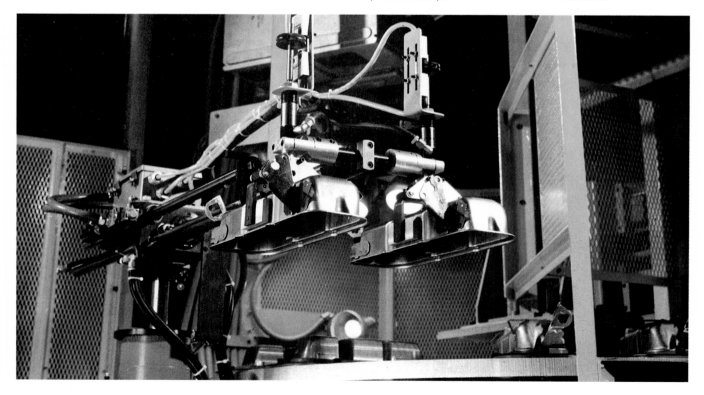

▼ To judge whether a room is dark enough to switch on the light, a robot needs a light meter controlled by a computer, instead of eyes and a brain.

▶ To weld something accurately, a sense of touch is needed. Here a welding robot senses the position of a piece of metal with pressure sensors.

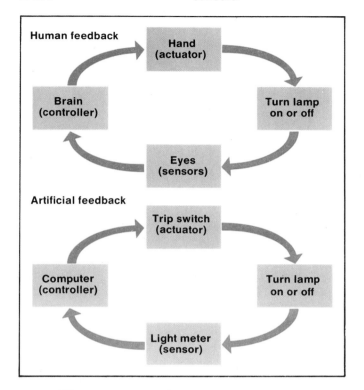

Human feedback

Hand (actuator) → Turn lamp on or off → Eyes (sensors) → Brain (controller) → Hand (actuator)

Artificial feedback

Trip switch (actuator) → Turn lamp on or off → Light meter (sensor) → Computer (controller) → Trip switch (actuator)

For example, when the robot bangs into a wall the force-sensors send that information to the robot "brain" which then tells the robot to change its actions so it doesn't bang into the wall again.

arm. The computer has been programmed to deal with, or process, the information supplied by the television camera. It changes it into a form which can be used by the computer controlling the robot arm, where it is added to other information already stored there. The robot now has enough information to carry out its job: instructions on what to do and a picture of the objects and tools it is working with.

One obvious advantage of a robot with vision is that it can be fed a series of differently shaped parts and handle each of them perfectly well. It will recognize each one and its "hand" will know the best place to pick it up. A "blind" robot would try to pick each one up in the same way, possibly with disastrous results.

Another way in which robots can have sense is through touch. Engineers can fix force-sensors to the end of a robot's hand. When these sensors touch something, information about what is happening is sent back to the robot's computer.

Feedback

The examples of vision and touch-sensitive robots demonstrate the important idea of "feedback". "Feedback" is the way in which the actions of people or things may be altered or changed by outside events. People use feedback all the time in their day-to-day lives. It comes into play, for instance, when you walk down a road, see a tree in your path and step to one side to avoid colliding with it and injuring yourself. Attempts to give robots feedback are necessary if they are to become useful in several different situations. Unfortunately for engineers, however, equipping robots with this ability is not easy. It is very difficult to code information from the outside world in a way that makes sense to the robot's computer, and then to make sure it processes the data quickly enough for the robot to respond. At present, probably only five per cent of the world's robots have sense. They are known as second-generation robots.

Robots on the Move

Armed with sensors, robots could have a great many uses in areas outside general industry. A popular place for them might be in the home. A Japanese company is spending about $300,000 a year on research to make home robots feasible. Such robots could, for instance, open the door when you arrive home, bring you your slippers and then pour you a drink.

Robots on Wheels

Robots would have to be mobile, as well as having sensors, to be able to find their way around. In Britain, engineering groups at GEC in Rugby and at the University of Warwick are trying to build robots on wheels. These machines are designed for jobs in factories, but they could be just as useful in the home. They would need a source of movement such as an electric motor, and a feedback system to send information about the outside world back to a steering mechanism.

Mobile robots would have a lot of other uses. For instance, they could work on building sites and down coal mines. For many years, engineers have dreamed of putting robots to work in unpleasant places such as mines. Sensors would guide them along tunnels to the coalface. At the coalface, the robots would cut the coal and hoist it on to trucks.

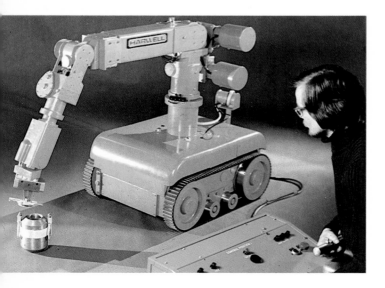

Robots on the Seabed

A still more promising place for robots could be under the sea. The bottom of the Pacific and other deep oceans is littered with potato-shaped lumps of rock that contain rare metals such as manganese and nickel. In a world that is gradually running out of its most precious metals, engineers have been concerned with finding a way to bring these lumps to the surface. Conventional methods would drag a series of sleds along the bottom of the sea or suck up the rock with a machine resembling a giant vacuum cleaner. But robots could actually crawl over the seabed looking for the most promising lumps. They would then load them into buckets which would carry them to the surface. The robots would have to find their way around sea-bed objects. They might even have to swim and take avoiding action if pursued by sharks. A coat of rustproof paint would also be necessary.

"Insect" Robots

Most mobile robots use wheels to get around, but some engineers are studying the possibilities of robots on legs. Although they look more like people, two-legged robots are frowned upon in engineering circles. Research workers have found it difficult to make robots of this kind move around easily. They are ungainly and keep falling down. Robots with six or even eight legs are much more practical. There have even been hints that robots of this kind are being developed in the Soviet Union for use in a possible war. Such "insect" robots would be very good at traveling over rough ground where ordinary vehicles, such as tanks, are unsuitable.

◄ Robots that can move have been the goal of engineers for years. In an early design, the UK Atomic Energy Authority devised a rudimentary robot, shown here, that moved around on caterpillar treads.

► Mobile robots of the future may operate under the sea (top) and down mines (bottom). The "swimming" robots would pick up metallic nodules from the sea-bed, or perhaps bring shipwrecks to the surface.

Robots in Space

Scientists have already made great use of machines in space. Although many people would love to copy astronauts and take a trip into the heavens, the truth is that space is not a very comfortable place. It is cold and dark and lacks all the comforts that make life pleasant, or at the very least bearable, on Earth. Still more important, the air we need to breathe to stay alive is missing in space.

Machines have enabled people to make use of space. Communications companies have put satellites into orbit. These send information from one part of the Earth to another. Other spacecraft send us information about the weather and all sorts of scientific data. For example, they send us information about the nature of the Solar System, and pictures of other planets.

Viking and *Voyager*

Most of this space hardware is definitely not in the robot category. It ranges from automatic machinery that regularly sends scientific data to Earth, to equipment which connects telephone calls without a human intervening. A few spacecraft, however, have come near to being robots. These include the *Viking* landers that the United States sent to Mars in 1976. They had scoop-like devices that dug into the soil to take samples. These were controlled by computers on board the landing craft. The latest *Voyager* spacecraft, which the United States sent deep into the Solar System in 1977, contain computers. These can guide the craft in a particular direction. The computers can be programmed by radio waves sent from Earth.

Space-farer Robots

Far more familiar robots may appear in space over the next few years. They would be more advanced than the robots that already toil away in factories on Earth. One exciting possibility for robots is in tending space factories of the kind that may become established around the end of this century.

The different characteristics of space (low gravity, no air, and abundant energy from the Sun) make it possible that some industrial processes could take place there under far more favorable conditions than on Earth. People would probably not want to work in factories out in space because of the unpleasant living conditions. It would be like working on an offshore oil rig, only far worse and much farther from home. These problems have led to the idea of recruiting a new breed of robot space-farers.

These robots might be fitted with little rocket motors so they could whizz their way around. They would also need suction grippers to walk around in a low-gravity environment instead of falling off into space. As they become more advanced and more skillful, the space-robots could work on other jobs – mending faulty spacecraft, for instance. They could also build huge platforms to house large panels of solar cells that would capture the Sun's energy and send it to Earth as microwaves. This would make a big contribution to solving the world's energy problems. Mobile robots on wheels could also trundle over the surface of planets, acting out the part of explorers.

Putting robots to work in space, instead of people, will also save money. According to calculations made recently, keeping a person in space for one hour and making sure he or she comes back to Earth alive costs about $15,000. Space robots would be expensive to develop, but once in orbit they would cost far less than humans to keep in operation. Furthermore, if a space-robot should break down, it could easily be replaced.

◀ The American space probe *Voyager* has begun a tour of the outer Solar System. A high point will be when it approaches Neptune in 1989. The probe is a prototype of the completely independent space robots we may see in the future.

▲ American space engineers have designed the Mars Rover, a robotic vehicle that could trundle around Mars. The vehicle would have sensors, such as TV cameras, to pick up information from the surroundings.

▶ The USA landed two *Viking* probes on Mars in 1976. They were able to analyse soil samples and the Martian atmosphere.

Robots in Fiction

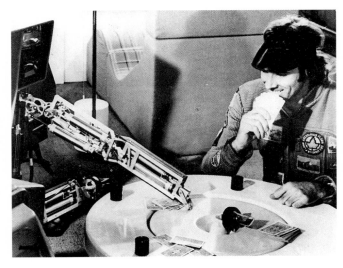

◄ The robot Gort, a 12-foot giant, appeared in *The Day the Earth Stood Still* (1951).

▼ Tobor from *Tobor the Great* (1954), was a robot hero who took care of his inventor's grandson.

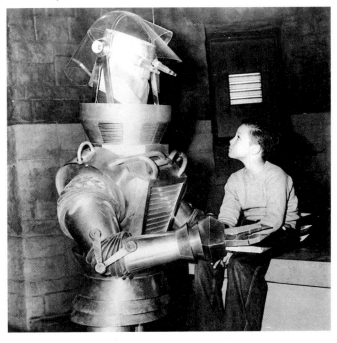

◄ In a film called *Silent Running* (1972), a robot demonstrates its ability to play cards with a human.

► The film *Star Wars* (1977) showed robots as friendly, lovable creatures. Shown here are C-3PO and R2-D2.

The Meaning of "Robot"

The word "robot" has its roots in fiction. The term came into the English language from the Czech word *robota* meaning "servitude". It is widely used mainly as a result of a play, by the Czech writer Karek Capek, called *Rossum's Universal Robots*, or *R.U.R.* The play was written in 1920 and translated into English three years later. It was first performed at the National Theater in Prague in 1921.

Rossum's Universal Robots

It is worth knowing something about *R.U.R.* because it set the tone for future plays and films about robots. In *R.U.R.* a factory, run by a man called Rossum, makes android machines that actually look and behave like people. The androids have a great capacity for work. At first Rossum sells plenty of them to other factories. But after a while events take a sinister turn. The robots learn how to think for themselves. They rebel against their human controllers and gradually take over the world.

This kind of turn-around appears in much of the fiction that followed *R.U.R.*, and many stories portrayed robots as evil and threatening. Even the words that Capek gave to Radius, the chief robot, are similar to those that following generations of robots have spoken as they dominate the cowering humans. By gaining possession of the factory, Radius explains in one of his speeches, "We have become masters of everything. The power of man has fallen. A new world has risen. The rule of the Robot."

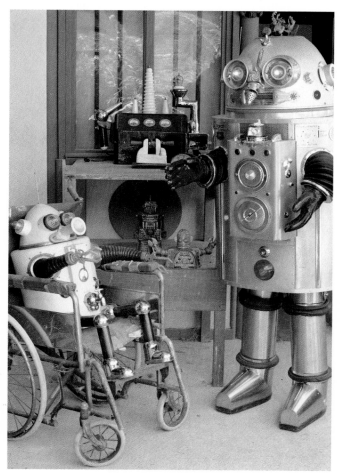

The Daleks, from the TV series *Dr Who*, are metal-clad visitors from outer space who wage war against humans.

Robots have also featured in exhibitions. Here one called ON/OFF, and a companion, go through their paces.

Robots in Films

Robots, mostly androids, also feature widely in films and on television. In *Target Earth*, a film made in 1944, the story is familiar. Alien robots from another planet attempt to take over the world. *Zombies of the Stratosphere* (1935) and *Phantom Empire* (1952) are both based on adventures in space. The robots in these films are also threatening. It was left to the film *Star Wars* (1977) to portray robots as friendly, even lovable, creatures that actually help people rather than try to kill or suppress them.

The "Three Laws of Robotics"

The writer Isaac Asimov has played a major part in presenting robots as beings with whom humans can find some kind of fellow feeling. In some of his stories the robots are made by a company called Robot and Mechanical Men Inc. This company employs a robot psychologist to make sure that each machine's brain works in such a way that the robot will not cause problems to people. Asimov even went to the trouble in 1940 of devising his famous 'three laws of robotics'. These are a sort of code of conduct designed to ensure that robots will act for the good of men and women.

The laws are:

1. A robot must not harm a human being, nor through inaction allow one to come to harm.
2. A robot must always obey human beings, unless that is in conflict with the first law.
3. A robot must protect itself from harm, unless this conflicts with laws 1 and 2.

Intelligent Robots

None of the robots we have looked at so far matches up to humans in one important respect – thinking. Even "second-generation" robots, the ones with sensors picking up information about the outside world, come a long way behind people in their ability to make sense of this information and act accordingly. "Second-generation" machines react simply as a reflex to what is taking place around them. The ability to weigh up the situation, perhaps comparing it with previous experience, is beyond them.

◀ A robot with "intelligence" is being designed by the US Navy. It will be able to make some decisions on the basis of information that it acquires from its surroundings. The machine will be able to swim under water and repair pipelines.

Third-generation Robots

But how do engineers propose to make the new style of computer? Two ideas are at the center of this work. First, researchers would like to change the way computers process their instructions or programs. At present, nearly all the computers in the world handle instructions one after another, or in sequence. This is because computers' electronic units are not very complex. Their layout is very simple when compared with the human brain.

Artificial Intelligence

"Thought", then, involves a rather more complex method of processing data than is possible with the computer "brains" of today's robots. The answer is to improve these computers. This brings us to an important area of research called artificial intelligence, or AI.

With AI techniques, computers could become far more powerful and versatile than they are today. They would be able to make decisions and so could be excellent substitutes for human workers. The new, AI-assisted computers would also act as the brains for a new breed of third-generation "intelligent" robots. Such machines would resemble the androids of science fiction because they would actually think like people.

In the human brain many lines of thought, or processing of ideas, can take place at once – in parallel. This is why humans can think of many unrelated ideas at once. Some people are able to connect these ideas to arrive at startling conclusions or to produce brilliant pieces of work. This is how, for instance, Sir Isaac Newton was able to devise his laws of gravity and Bach to compose his music.

Designing machines with this element of parallel processing is not easy. Different strands of "thought" patterns must be set in motion and then linked up with each other again from time to time. Making a machine that works like this is rather like planning a complicated railroad network. Such a network must make a large number of connections and every train must be on time – down to the nearest split second.

The second way to make intelligent computers is by redesigning the memories. Where today's computers have one memory, the new machines would have two. One would store programs and data relevant to the job in hand and the other would be a "knowledge base". It would store the great mass of general knowledge that makes it possible for humans to move around and function normally.

Linking up the "knowledge base" with the computer memory and the machine's central processing unit is necessary if "third-generation" robots are to become reality. Researchers are already working on these problems.

▶ Factory robots could be of greater use if they were mobile. For instance, robots of this kind could trundle around machine-shop floors repairing tools or delivering goods. Engineers are developing such devices. These could be on sale in the late 1980s.

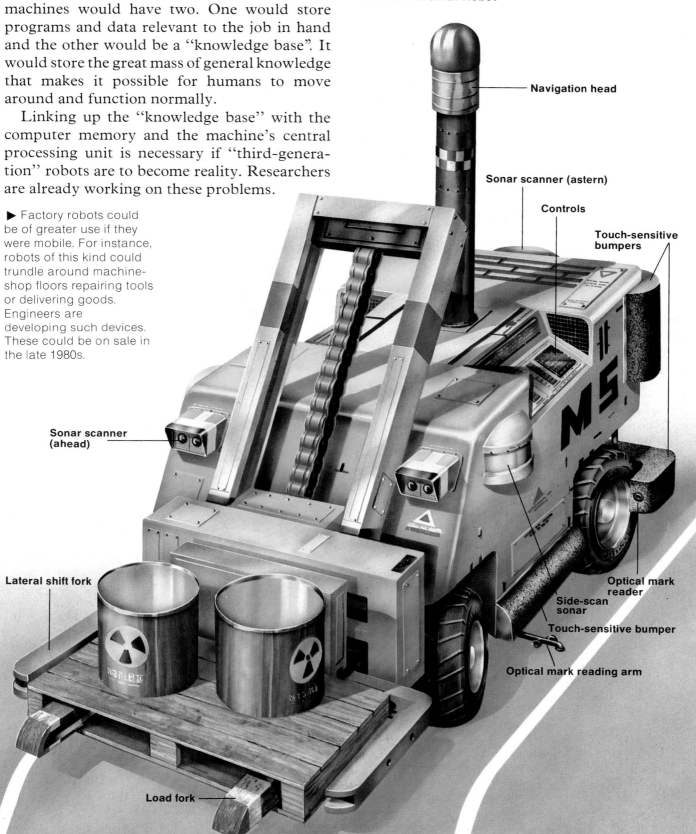

Mobile Industrial Robot

Navigation head

Sonar scanner (astern)

Controls

Touch-sensitive bumpers

Sonar scanner (ahead)

Optical mark reader

Side-scan sonar

Touch-sensitive bumper

Optical mark reading arm

Lateral shift fork

Load fork

Robots Built by Robots

In the most terrifying science fiction stories involving robots, they first take over the world and enslave the human population. Then, and this is the point at which shivers start to crawl down people's spines, the robots start to breed. Although robots that can reproduce themselves have not yet appeared in real life, researchers in the United States have recently laid down the engineering guidelines which they would probably follow.

John von Neumann, a Hungarian-American scientist who also did much of the early work on computers, started it all in the 1940s. He proposed that machines should be able to reproduce by following a given set of rules. Scientists later found out that these rules are remarkably similar to the way that animal cells reproduce themselves.

According to von Neumann's theory, a self-reproducing robot system would need four parts. The first is an automatic factory which collects raw materials and turns them into products according to written instructions. The second part is a duplicator which takes a written instruction and copies it. The third part is a controller which sends the instruction to the duplicator for copying. One copy is passed to the factory for action and the other to the final parts produced by the factory. The final part of the system is the written instruction itself which tells the factory to construct a complete new automatic factory out of the parts it has made.

Factories on the Moon

Scientists played around with these ideas for years but then, in the summer of 1980, researchers employed by the National Aeronautics and Space Administration (NASA) produced an engineering blueprint of how a self-replicating system would operate. The way it works is shown in the illustration.

The system has four main parts. There is a processing unit which mines materials and refines them. There is also a production system which makes parts from what the processing unit provides. Another production system turns out complete products from individual components. Finally there is a "universal constructor". This constructor is the key to the whole system. It takes parts from the production systems and builds complete new factories with them. As the new factories contain their own universal constructors, the whole process

can begin all over again. In this way the factories can reproduce themselves at a fast rate.

In each universal constructor is a series of robots. These know how to make parts for the factories themselves, but they are also intelligent enough to turn out new robots. An interesting point about the NASA exercise is that it is hoped to put the system to work on the Moon one day. Such a space factory would make great use of the valuable raw materials on the Moon, and it would never need to be serviced from Earth. It could continually provide itself with a new supply of robot workers.

▼ People will probably return to the Moon in the 1990s or early next century to mine raw materials. The moon factories will quite likely be run by robots. The robots may self-replicate so that the factories have an expanding workforce

The Robot Future

Since factories were first built, people have been eager to invent new kinds of machinery that will do the jobs of factory workers. As industrialization increased during the 19th century, factory owners found they could produce more goods, not by taking on more workers, but by installing new machinery.

At the beginning of the 18th century in Britain, about 80 per cent of the population of 5 million worked on the land. By the middle of the 19th century half the country's working population had jobs in factories. From the beginning of the 20th century this proportion has steadily fallen to about 35 per cent today. Other industrialized nations throughout the world have followed similar patterns.

Robots – A Mixed Blessing

The widespread use of automated machinery, computers and robots will continue this trend. Fewer and fewer jobs in manufacturing industries will be available to people wanting work. Many will find jobs in other industries that are growing, such as the "service" industries – catering, local government, banking and transport. But just what will happen is difficult to tell.

Employment in the Future

In the past, new types of employment appeared for workers whose jobs had disappeared. In the 19th century, for example, millions of jobs were created by the growth of factories. Yet, only a few years before, no one would even have been able to tell what went on in such strange places.

On the other hand, robots taking over jobs in factories could worsen the high unemployment rate that the industrialized world already suffers from. Although new industries (such as making and programming electronic goods) are being created, they require few workers. Many people are worried that not enough jobs will be created to absorb the people no longer needed in factories.

Learning to Live with Robots

The development of robots will have another, perhaps deeper, effect on society. Robots will make their presence felt not just in factories but in many other areas – in the home, for example, and in hospitals and other service jobs. People have had little trouble in getting used to new forms of electronic gadgetry such as washing-machines, home computers and video recorders. But they may find the new forms of robot more difficult to accept easily.

The "intelligent" robots discussed earlier may produce mixed feelings among people who work with or control them. Some writers argue that people will welcome machines they can get on with easily. There would be no need to address the new breed of robot with special computer languages. Most probably people would command them simply by talking to them. But, on the other hand, men and women may feel uneasy when confronted by machines that are as "clever" as, or cleverer than, themselves.

The Robots of the Future

One thing is sure. The robots that wander around homes and factories during the next century will not be the metal-clad, cumbersome monsters of past science fiction films. They will be specially designed to fit in with their surroundings. The robots of the future will probably be sleek and smooth, and look like ultra-efficient vacuum cleaners. Robots will gradually become as familiar as the electric motors that abound in homes and factories today.

▶ Autonomous machines will do many of the world's mundane jobs a century or two from now. It is impossible to predict how far into everyday life the machines will intrude. But undoubtedly the jobs that people find the most unpleasant will qualify first for the robot takeover. We may well see robot "life savers" rescuing people from floods, as shown here. Robot fire engines and robot lifeguards on the beach are further possibilities.

Glossary

AC Alternating current.

Acoustic coupler Equipment which turns computer signals into telephone signals and back again.

Acoustics The science of sound. Popularly refers to the acoustical character of halls and rooms.

Actuator A device responsible for action, e.g. robot hand, car wheel, person's foot.

AFC Automatic frequency control. A circuit that controls drift in FM tuners.

AM Amplitude modulation. A common form of radio broadcasting in which the carrier frequency is constant and the carrier amplitude is varied.

Ambisonics A system of recording and reproducing sound which gives 'all-round' sound, an improvement on stereo.

Amplifier An electronic device for magnifying, and usually controlling, electrical signals. Amplifiers originally used thermionic valves, but now transistors and integrated circuits are used.

Android A machine that both looks and behaves like a human being. The word 'android' comes from the Greek words *andros* ('man') and *eidos* ('form').

Antenna A device for intercepting radio signals to feed to a tuner or receiver. It may be a wire, ferrite rod or tuned dipole.

Artificial intelligence (AI) The discipline by which engineers try to make computers copy human thought-patterns. AI-assisted computers will be used as the 'brains' for future third-generation robots.

Audio frequency Frequency within the range of human hearing (20 to 20,000 Hertz).

Automatic machine Equipment that operates according to a fixed set of instructions. Not to be confused with robots. Automatic machines include devices such as clocks and traffic lights.

Automation The process of linking up computerized tools and machines and setting them to work with the minimum of human intervention.

Aux Auxiliary input, on an amplifier for example. It accepts an extra signal such as tape cassette output.

AVC Automatic volume control (in a radio set).

Baffle The panel on which a loudspeaker is mounted to isolate the front and rear of the diaphragm.

Balance Relative volume levels between the channels in a stereo sound system. Also the volume between high and low frequency ranges.

Barn doors Hinged black flaps attached to studio lights. They can be moved forward and backward to increase or decrease the amount of light.

Bass The low-frequency end of the sound range (below about 150 Hertz).

Bass reflex A type of loudspeaker cabinet with a duct in the speaker face to use the resonant effect of sound waves inside the box.

Binary Microcomputers use a two-state, or binary, code to represent the ons and offs of electrical voltages.

Binaural Two-channel sound in which each channel is heard only through one ear, as with headphones.

Bird Television slang for 'satellite'.

Bit Digital signal. It is derived from the words 'binary digits'.

Boolean Algebra The rules of logic used by microcomputers to take decisions based on the facts they have been given.

Booster station Television signals grow weaker as they travel away from transmitters. Booster stations pick them up, strengthen them and pass them on.

Bugs Mistakes in your program.

Byte Eight bits (usually enough space for a letter or number).

Cable TV A method of carrying pictures and sound to television sets by cable. The cable may be buried underground or strung on telegraph poles.

Carrier waves These are generated by TV engineers to carry sound and picture signals through the air.

Cartridge A record-player cartridge or pickup device which generates voltage to feed an amplifier.

Casseiver A three-in-one unit combining a radio tuner, amplifier and cassette deck.

Cathode ray tube A glass vacuum tube found in television sets. It has a broad end (the screen) and a narrow end containing electron guns.

CB Citizens' Band radio.

CD Compact Disk.

Central processing unit (CPU) The part of a computer that does all the information processing.

Channel Television signals travel on a limited number of frequencies through the air. Each TV channel is given a different frequency.

Channel programmer An electronic control unit on a video recorder. It enables the video recorder to stop, start and change channels on a time control.

Chrominance signal The color signal produced by a TV camera. It is made up from the primary colors of light: red, green and blue.

Cincinnati Milacron A popular form of industrial robot.

Closed-circuit TV A system for viewing without transmission being involved. A TV monitor, rather than a TV receiver, is used to display the pictures.

Coaxial cable A cable that consists of a tube of electrically conducting material, inside which is a central conductor. The central conductor is insulated from the tube. These cables are used to carry very high frequency signals such as those used in televison.

Color balance A color mechanism on a TV camera which enables it to be adjusted for lighting differences indoors and outdoors.

Compatible Several meanings: (1) An FM stereo radio signal receivable as mono by a radio tuner, (2) a stereo record playable with a mono pickup, (3) the ability to connect various units together in a system without an electrical mismatch.

Compiler An automatic translator that turns programs into the microcomputer's binary code.

Control unit The vital part of a robot that tells it what to do. This is normally a computer.

Crosstalk A breakthrough, or signal leakage, between two channels of a stereo system.

Data A collection of facts.

DBS Short for direct broadcast satellite. A DBS sends out a signal strong enough to be received by a domestic dish antenna.

DC Direct current.

Decibel (dB) A logarithmic unit of measurement used to show relative levels of current, voltage or sound power. Regarded as a ratio, these figures may be very big, say 10,000 or 100,000 times. The dB was introduced to give these levels by an easy-to-grasp figure. For example, a 10,000 to 1 ratio is 40 dB, and 1,000,000 is 60 dB. Doubling the power means a rise of 3 dB.

Degausser (Or defluxer) A device that removes unwanted magnetism in tape heads or on tapes.

Digitizer A device used for getting pictures into your microcomputer.

DIN Short for Deutsche Industrie Normen. A set of standards established in Germany for hi-fi equipment. Also covers plugs and connectors.

Disk Floppy disks and hard disks can be used for storing data when your computer's main memory is full.

Distortion Any difference in sound reproduction from the original source. Many forms of distortion arise in sound recording and replay.

Dolby sytem Noise-reducing circuits used to cut down hiss and noise in tape machines.

Driver An individual loudspeaker unit.

Dubbing Copying tapes or disks from previously recorded versions.

Earpiece A listening device worn by the presenter of a TV program. It is not visible to the audience but enables the director in the gallery to talk directly to the presenter.

Eastman Color 16-millimeter-wide film which produces a negative image from which prints are made for transmission.

Ektachrome The common name for 16-

millimeter-wide film which produces one positive image only.

Electron guns In a television set, electron guns shoot electrons through the slots in the shadow mask on to the phosphor dots on the screen. This makes the phosphor dots glow with color to produce a picture. In a television camera, the electron guns fire electrons at the target plate to scan the pattern of the image and turn it into electrical signals.

Electrostatic (ELS) Refers in hi-fi to loudspeakers and headphones that use fixed and movable plates, across which a high-voltage charge is applied.

ENG Short for electronic news gathering. The news team carry a portable electronic camera linked by cable to a video recorder.

Feedback The mechanism by which a sequence of events influences action. A good example of a device providing feedback is a thermostat. Thermostats respond to heat and so are often used to control the temperature of central-heating systems automatically.

Field Effect Transistor (FET) A special kind of transistor which amplifies voltage, not current.

Filter An electrical circuit that suppresses or reduces a certain part of the audio frequency range.

First-generation robot The most common kind of robot in which a series of instructions controls the robot's arm. There is nothing fancy about these robots; they lack any form of 'sense'.

Flats Movable pieces of scenery used in a TV studio.

Flexible manufacturing systems A series of robots and machine tools that turn out a lot of different types of parts under computer control.

Floor manager The human link between the TV gallery and the studio floor. The manager uses headphones to listen to instructions from the director in the gallery.

Flowchart A useful way of breaking a problem down into the steps a computer will have to follow to solve it.

Flutter Rapid variations in the speed of a tape mechanism or turntable. Causes a wavering of musical pitch.

FM A form of radio broadcasting in which the carrier amplitude is constant but the carrier frequency is varied. FM gives a wide range response and freedom from noise.

Gallery The control room at a TV studio which contains the mixing desk and monitors.

Graphics Drawing pictures and diagrams.

Hand The important part of a robot arm. No robot could work without one. Robot hands include suction pads, claws and grippers.

Hardware Everything you can see and touch about your microcomputer.

Hertz (Hz) Unit of measurement for sound frequency. A frequency of one Hertz equals one cycle (complete sound wave) per second. One Kilohertz (KHz) equals 1000 Hertz.

High-level languages   Programming languages that resemble ordinary English.

Hydraulic robot   A machine in which the hand and arm of the robot are moved by pressurized liquid acting on mechanical linkages.

Infinite baffle   A type of loudspeaker mounting in which the sound waves from the back of the cone are prevented from reaching the front.

Input   Putting things in to your microcomputer. Input devices are machines that feed information into your microcomputer.

Integrated circuit (IC)   A solid block containing the functions of numerous transistors and other electronic components.

Interface   If you want to connect two pieces of equipment together, they must use the same code. Another way of saying this is that the interface must match.

K   An abbreviation for kilo, which means a thousand. It usually stands for kilobyte when used in connection with computers. One kilobyte equals around a thousand bytes.

Keyboard   The most common input unit. It looks like a typewriter, but is much quieter in operation.

Knowledge base   A feature of advanced forms of computers which contains general data about the world and feeds this to the rest of the computer.

Lightpen   This looks like a pen but has a light-sensitive tip. It can be used with a screen to draw pictures.

LSI   Stands for Large Scale Integration, and is used in connection with silicon chips. It refers to the fact that there are a lot of tiny transistors built on to the surface of a chip. VLSI means Very Large Scale Integration.

Luminance signal   The black-and-white signal produced by a TV camera. Sometimes this is produced by a separate tube in the camera, sometimes by the three color tubes.

Machine code   The language that computers use. Compilers and translators turn program languages into machine code.

Machine tool   An industrial device that cuts, grinds, bores or otherwise fashions pieces of metal, wood or plastic. The latest machine tools are controlled by computers and are fed with parts by robots.

Mechanization   The first stage in the process by which machines take over the work of humans. This word first came into widespread use with the 'factory age' ushered in by the Industrial Revolution.

Main memory   Needed by the CPU to deal quickly with its immediate needs. The CPU of a microcomputer can't hold on to any information without the main memory.

Mega   Short for million. A megabyte, for example, is a million bytes.

Memory   The part of a computer used to hold or 'store' programs and data.

Microelectronics   The variety of products that come on silicon chips.

Microphone Used for transforming sounds into electrical signals. In this form, sound signals can be recorded and then played back through a loudspeaker.

Microprocessor A type of silicon chip built to act as a CPU.

Millimeter The traditional unit of measurement for film width. Film used for recording TV programs is usually 16 millimeters wide.

Mixing desk A control center into which different picture sources can be fed for displaying on monitors. The director selects each picture for transmission and changes it when he or she likes.

Modem Similar to an acoustic coupler, but more sophisticated.

Modulation The control of one signal by another, by altering the amplitude or frequency of the 'carrier' signal.

Monitor A screen used for closed-circuit picture display. Monitors, unlike TV sets used in the home, do not receive broadcast pictures via an antenna.

Numerical control The process by which machine tools or robots are controlled by binary digits.

Operating system Software which tells a microcomputer how to work. Also known as systems software. It is often programmed into a ROM chip.

Optical-fiber cable Thin glass fibers bunched together to make a cable capable of carrying a very large number of TV channels. TV signals travel along the cable as pulses of light. They can travel long distances before they need boosting.

Output Getting things out of your microcomputer. Output can include words on paper, pictures on a screen, and even sound.

Outside broadcasts TV programs made outside a television studio. Outside broadcast programs can be fed into the network by using a portable transmitting antenna.

Packages Ready-made software programs.

Peripheral Any device that you can connect to your microcomputer.

Phono plug A coaxial plug used to connect hi-fi equipment.

Phosphor dots Used to coat the inside of a TV screen. The phosphor dots glow with color to form a picture when they are hit by a stream of electrons.

Pneumatic robot A robot which is driven by the movement of compressed air.

Presenter A person who, in many television programs, appears on the screen to introduce various items and guests. He or she is often called the frontman, or linkman.

Printers Devices used for getting your output on to paper as words or numbers.

Program Software instructions that tell a microcomputer how to solve a problem.

Programmer Someone who writes computer programs.

RAM Short for Random-Access Memory. It is a type of chip used for main memories. A 1K RAM holds one kilobyte of main memory.

Rigger driver The person who drives the outside broadcast vehicle (the scanner) and does many of the hard jobs involved in setting up an outside broadcast.

Robot The modern definition is a mechanical arm controlled by a computer so it can copy the actions of humans in a flexible, universal way. However, the term can loosely be taken to mean any machine which copies human beings in appearance or action.

ROM Read-Only Memory. A ROM chip holds information, such as operating systems, which does not need to be changed.

Satellite Communications satellites pick up TV and telephone signals beamed up from Earth, amplify them, and transmit them to receiving stations thousands of miles away from the transmitting stations.

Second-generation robot A robot with one or more sensors.

Semiconductor A material which is normally a poor conductor of electricity but, when treated, allows current to flow. Silicon is the most common semiconductor used in micro-electronics.

Sensor A device that feeds information from the surroundings to a person or device. Important human sensors are the eyes, ears, nose and nerve endings. Common types of sensors in robots are TV cameras and force- or pressure-sensors.

Silicon chips Tiny pieces of silicon designed to hold the electronic circuits that store and process the information passing through your microcomputer.

Software The ideas and instructions that tell your microcomputer what to do. Software is built up of programs.

Solid state Circuits using semiconductors, such as transistors and integrated circuits.

Stereophonic (stereo) Sound recorded and replayed from two or more channels to give the listener an impression of where the sounds are coming from in space.

Tape Used as a storage device with microcomputers. Information is recorded on the magnetic surface of the tape.

Teaching by doing A method of instructing robots in which a human operator takes the robot through a particular action. The robot remembers the action for later use.

Telechiric arm Mechanical arm controlled continuously by a human operator. Not to be confused with robots, which are controlled by computers.

Terminal Unit connected to a computer to feed in, or gain access to, information. Terminals include keyboards and visual display units (screens).

Third-generation robot A robot with artificial intelligence.

Three-dimensional TV A way of making an image more life-like by showing it from two different directions and superimposing the two pictures.

Tracking Moving a TV camera forward, backward or sideways without using the zoom.

Transducer A device for converting one form of energy to another, e.g. loudspeakers.

Transmitter An engineering center with a tall mast which is used to send out television signals through the air.

Treble The upper end of the sound range.

Tuner The mechanism inside a TV set which is used to select different signals and feed them to the channel buttons on the set.

Unimate A popular form of robot made by the American firm Unimation.

Unmanned factory An advanced form of factory. In reality a few people are required for jobs such as administration, maintenance, and controlling the computers. Unmanned plants will most probably become widespread in the 21st century.

User group A group of people interested in microcomputers (who usually use the same type of machine). If you write to your computer manufacturer, they will give you the address of your local user group. If there isn't one, then start your own.

Very large scale integration The process of packing many electronic components into a small space to make powerful computers.

VHF Very high frequency. Radio frequencies between 30 and 300 MHz.

Video cassette An enclosed reel of magnetic tape for recording on commercial or home video recorders. The tape carries pictures and sound.

Video disk A flat disk on which pictures and sound are recorded. It is replayed on a video-disk player.

Viewdata System which connects ordinary television sets to computers via telephone lines to gain access to information.

Vision mixer The person who sits at the mixing desk in the gallery next to the director. The vision mixer operates the switches controlling the picture sources.

Vision robot A robot with a television camera to enable it to 'see' what it is doing.

Visual Display Unit (VDU) A cathode ray tube, as found in a TV set, for displaying the output of a computer.

Word The number of bits (binary 0s or 1s) your microcomputer can cope with at once depends on its word length. This is often eight bits, which equal one byte.

Wow and flutter Unwanted changes in disk or tape speed. "Wow" refers to slow speed, and flutter refers to changes above about 10 Hz.

XY co-ordinates When you want to describe a particular spot on a screen, or on a sheet of paper, it is common to talk about x and y co-ordinates. The x counts the points across the screen or sheet, and the y counts the points down. Where the two meet is the point being described.

Zooming Closing in to, or pulling out from, a subject without moving the TV camera. This is done with the aid of a zoom lens.

Index

ACKNOWLEDGEMENTS

Page 19: Intel Corp; 21: Plessey Semiconductors Ltd; 31: Hewlett Packard Ltd; 32: Atari; 36 top: Commodore Computers Ltd; 36 bottom: Atari; 37 top: Redifusion Computers Ltd; 37 bottom: Atari; 46–47: SFP/Jean-Claude Pierdet; 48: TVS; 49: ITN; 50: SFP/Jean-Claude Pierdet; 52–53: TVS; 56 left: Gretag Elektronik GMBH; 64 top: Metropolitan Police, New Scotland Yard; 64 bottom: UK Atomic Energy Authority; 65 top: ZEFA; 65 bottom: Sony (UK) Ltd; 84: David Redfern; 85 top: John Price Studios/EMI; 85 bottom: SARM Studios; 88: BBC; 89: BBC; 92 British Telecom; 93 J. Allan Cash; 100: Cincinnati Milacron; 101 left: Trallfa Nils Underhang A/S; 101 right: British Robot Association; 102: UK Atomic Energy Authority; 104: British Leyland; 107: Devilbiss Co. Ltd; 108: Autoplace Inc./Visual Arts; 109: Hitachi Ltd; 110: UK Atomic Energy Authority; 112–113: NASA; 114: National Film Archive; 115 bottom left: Courtesy of Lucasfilm UK Ltd; 115 right: Clayton Bailey; 115 top left: BBC Copyright Photograph; 116: Dept of the United States Navy.

Picture research: Penny Warn and Jackie Cookson

The publishers would like to thank the following for their help in the preparation of this book: Alan Rudge, Department of Design Research, Royal College of Art; Alphatronic Microprocessor Applications Ltd; Atari International Inc; David Gwyn Jones; Roger Andrew Mintern; Mullard Limited; Philips Electronics; British Robot Association; Taylor Hitec Ltd; University of Warwick Robot Laboratory.